Ultrasonic Tomography Approaches and Applications

Ultrasonic Tomography Approaches and Applications details cutting-edge guided wave tomographic techniques and their applications in the non-destructive evaluation and structural health monitoring industry. Using technology which has the power to significantly transform structural maintenance and integrity management, the book enables readers to apply these techniques through the use of case studies and practical end of chapter problems.

Ultrasonic guided wave tomography (UGWT) delivers accurate and robust results to diagnose defects and inclusions in thin sheets of many different materials. Applicable to many different industries, the book uses quantitative guided wave imaging approaches based on diffraction tomography (DT) and full waveform inversion (FWI), comparing and contrasting their usefulness in reconstructing physical properties in isotropic and anisotropic structures. Applying this to real-world case studies such as liquid-loaded tank floors and pipeline corrosion monitoring systems, the book demonstrates how UGWT is revolutionizing non-destructive evaluation and structural health monitoring. It presents the theoretical analysis, simulations, experiments, and practical applications of the UGWT method, ensuring more reliable non-destructive testing can be carried out.

This book will be of interest to students and engineers in the field of ultrasonics, maintenance engineering, mechanical and structural engineering, acoustics, vibration, and materials science

To assist students and instructors, the book comes with online resources like figure slides and solution manual.

Ultrasonic Tomography Approaches and Applications

Jing Rao and Madis Ratassepp

CRC Press
Taylor & Francis Group
Boca Raton London New York

CRC Press is an imprint of the
Taylor & Francis Group, an **informa** business

Designed cover image: Jing Rao and Madis Ratassepp

First edition published 2026
by CRC Press
2385 NW Executive Center Drive, Suite 320, Boca Raton FL 33431

and by CRC Press
4 Park Square, Milton Park, Abingdon, Oxon, OX14 4RN

CRC Press is an imprint of Taylor & Francis Group, LLC

ISBN: 978-1-032-76640-9 (hbk)
ISBN: 978-1-041-13065-9 (pbk)
ISBN: 978-1-003-66792-6 (ebk)

DOI: 10.1201/9781003667926

Typeset in Nimbus Roman font
by KnowledgeWorks Global Ltd.

Access the Instructor and Student Resources:www.routledge.com/9781032766409

Dedication

To my family and my supervisors

Contents

Preface

Ultrasonic non-destructive testing (NDT) & evaluation (NDE) and structural health monitoring (SHM) of materials and structures have been widely used in aerospace, civil, nuclear, and oil & gas industries. Traditional ultrasonic NDT & NDE and SHM techniques are mainly used to determine the location and size of defects without providing physical properties such as velocity and density. With the dramatic increase in the use of high-performance engineered materials and components, more detailed information on physical properties is required to assess the failure probability of structures used in these applications. Ultrasonic bulk/guided wave tomography is a promising tool to obtain quantitative information on physical properties.

In ultrasonic bulk/guided wave tomography, there are two major steps to be considered: first, quantitative information is extracted from ultrasonic bulk/guided waveforms by different techniques; second, this information must be appropriately processed to produce the mapping of the object parameter by using tomography methods. The purpose of this book is to bring readers up to speed with the latest developments and applications of ultrasonic tomography methods. This book is designed for students, researchers, and engineers familiar with this field to serve as a basis for their work. This book can provide the understanding and application of quantitative ultrasonic tomography methods. These quantitative ultrasonic imaging methods can achieve a step change in the NDT & NDE and SHM of materials and structures used in various industries. The core content of this book is ultrasonic bulk/guided wave tomography and its applications, including analysis of reconstruction resolution and accuracy, limited view problems, anisotropic effects, etc.

The book begins with an overview and background materials in Chapter 1 and then continues to introduce different tomography methods commonly used in Chapters 2 and 3. Ultrasonic guided wave tomography based on mono-parameter full waveform inversion (FWI) and its applications are introduced in Chapters 4 through 8. In Chapter 9, the improved mono-parameter gradient sampling FWI is presented, and finally, the ultrasonic bulk wave tomography based on multi-parameter FWI is described in Chapter 10.

The authors have had the good fortune to witness the rapid development of ultrasonic bulk/guided waves in NDT & NDE and SHM. They are very interested in the reconstruction resolution, accuracy, and reliability of the ultrasonic bulk/guided wave tomography methods. Ultrasonic bulk/guided waves can be used to non-destructively evaluate physical properties and to detect and characterize subsurface defects in critical structures. This work provides a valuable quality control tool for equipment and in-service inspection procedures for structures such as pipelines, bridges, aircraft, automobiles, and nuclear power components, whose integrity is critical to public safety.

Author Biographies

Jing Rao is a Professor at Beihang University. She previously served as a Lecturer at the University of New South Wales (UNSW), Australia, and held an Alexander von Humboldt Fellowship at the Technical University of Munich, Germany. She received her Ph.D. degree from Nanyang Technological University, Singapore. Her research focuses on ultrasonic non-destructive testing and structural health monitoring. She has led over 30 research projects funded by the National Natural Science Foundation of China, the Beijing Natural Science Foundation, the Alexander von Humboldt Foundation, and UNSW Australia, among others. Currently, she serves as an Associate Editor for *Measurement*, *Ultrasonic Imaging*, and the *IEEE Open Journal of Signal Processing*. Her recent conference roles include serving as a Session Chair for the 2021 Australian Aerospace and Applied Mechanics Congress and the 2023 International Congress on Ultrasonics, as well as a Technical Program Committee Member for the 2023 IEEE International Ultrasonics Symposium. She is also a committee member of the Structural Health Monitoring and Prognostics Division of the China Instrument and Control Society and the Non-Destructive Testing Division of the Chinese Mechanical Engineering Society.

Madis Ratassepp is a Senior Research Fellow at Tallinn University of Technology (TalTech), Estonia, specializing in wave propagation modeling and ultrasonic non-destructive evaluation. He received his Ph.D. degree in Civil and Environmental Engineering from TalTech. His research focuses on guided wave tomography, full waveform inversion, and advanced signal processing for structural health monitoring of composite structures, pipes, and plates. He has led several key research projects, including *Quantitative Imaging Methods and Efficient Signal Processing Algorithms for Ultrasonic Non-Destructive Evaluation* (2020–2024), and currently serves as Principal Investigator of *Advanced Acoustic Sensing and AI-Driven Underwater Monitoring Methods for Maritime Environmental Protection and Security* (2026–2030). His work has been published in leading journals such as *Ultrasonics*, *NDT & E International*, and *Mechanical Systems and Signal Processing*. Dr. Ratassepp supervises Ph.D. and Master's students, actively serves as a reviewer for international journals, and holds academic leadership roles including Doctoral Program Manager in Civil Engineering and Architecture at TalTech.

Acknowledgments

The authors would like to thank Ernst Rank (Technical University of Munich, Germany), Zheng Fan (Nanyang Technological University, Singapore), Ching-Tai Ng (Adelaide University, Australia), Carlos Omar Rasgado Moreno (Testia GmbH, Germany), Marek Rist, Raul Land, and Paul Annus (Tallinn University of Technology, Estonia), and Junfeng Liu (Beihang University, China) for their valuable contributions, discussions, and support throughout the development of this book.

1 Introduction

1.1 BACKGROUND

Quantitative imaging can provide quantitative estimations of physical properties such as velocity and density. The ability to reconstruct subsurface physical properties from ultrasonic wave measurements is very valuable in many applications as it may enable non-destructive testing (NDT) & evaluation (NDE) and structural health monitoring (SHM) approaches with high sensitivity. For example, numerous defects are reflected in ultrasonic velocity variations, which are different from the background ultrasonic velocity, so velocity reconstruction from wave measurements can be used for defect detection. Based on this concept, ultrasonic tomography provides a promising alternative for defect diagnosis in a variety of materials. In this book, ultrasonic tomography includes ultrasonic guided wave tomography (UGWT) and ultrasonic bulk wave tomography (UBWT). UGWT is generally aimed at thin sheets, and UBWT can detect deep defects in thick materials, which will be fully explained later in this chapter. Also, UGWT can enable the detection of inaccessible areas (e.g., pipe supports or under clamps) that present significant challenges to conventional ultrasonic NDT & NDE methods. The field of ultrasonic tomography has attracted a great deal of interest and has become a crucial subject in NDT & NDE and SHM. There has been a significant increase in the number of publications, research activities, and in-service field testing applications.

The key to realizing this potential in ultrasonic tomography is the applicability and usefulness of accurate, robust, and high-resolution algorithms for reconstructions of ultrasonic wave velocity. Tomography algorithms are used to convert the measured signals into the mapping of physical properties. On this basis, for example, velocity imaging can be further transformed into thickness imaging based on dispersion curves in UGWT. The quantitative reconstruction of the remaining wall thickness is critical for determining the remaining life of structures. Although quantitative tomographic approaches for ultrasonic tomography show the capability to reconstruct defects, they are relatively underdeveloped as the majority of tomography applications are based on ray assumptions. These assumptions ignore diffraction and thus severely reduce the reconstruction resolution. Only large and smoothly varying defects in materials and structures can be reconstructed well. This study has aroused considerable interest among researchers, engineers, and scientists, who have further developed different ultrasonic imaging methods to improve accuracy, resolution, imaging time, etc. The knowledge presented in this book will generate creative ideas that can be used in developing new inspection techniques and applications.

Ideally, the quantitative reconstruction can be obtained using different tomographic algorithms when the measurements are conducted from all possible angles around the inspection area—a full view configuration. However, in many real applications, only a limited range of viewing angles is available. Also, most of the existing

DOI: 10.1201/9781003667926-1

work on UGWT focuses on waveguides with free boundaries. However, it remains challenging to apply UGWT to waveguides with liquid loading, as the attenuation of the guided waves leads to a low signal-to-noise ratio (SNR). Besides, most ultrasonic tomography methods being based on the assumption of isotropic material models, there are many other engineering applications that make use of anisotropic materials. In addition, feedback from field experiences and actual encounters leads to the development of many new problem statements and considerations, effectively addressing these challenges. The work presented in this book serves as a valuable starting point. Hundreds of papers, patents, and other related works prove extremely useful in addressing current challenges. Furthermore, breakthroughs in ultrasonic tomography methods and applications are expected to continue. Ultrasonic tomography will be intricately linked to advancements in sensor development, signal analysis and processing, defect assessment, and prediction of remaining useful life, among others. These advancements will enhance the safety, robustness, reliability, and cost-effectiveness of inspections through self-diagnosis in SHM.

1.2 ULTRASONIC BULK WAVES AND GUIDED WAVES

Ultrasonic waves have been widely used in NDT & NDE, SHM, medical imaging, cleaning and surface treatment, sonar and underwater applications, and distance measurement. This book focuses on the use of ultrasonic waves for NDT & NDE and SHM. Compared to other NDT & NDE techniques, the advantages of ultrasonic techniques include sensitivity to surface and subsurface defects, high accuracy in determining the defect locations, sizes, and shapes, and the ability to enable large-area inspection of the test object. However, ultrasonic NDT & NDE also has its disadvantages. For instance, measured ultrasonic signals must be inspected by experienced technicians, and unless non-contact methods are used, couplants are required to provide efficient transmission of ultrasonic energy between the test objects and the transducers.

1.2.1 ULTRASONIC BULK WAVES

The propagation of elastic waves in various media, such as a semi-infinite solid medium, an anisotropic medium, and a multi-layered medium, has attracted extensive research. For the sake of simplicity, this section focuses on elastic wave propagation in an infinite, isotropic, and homogeneous medium. The primary equations are provided here, and more details can be found in [12, 26, 197]. In the absence of an external force, the equation of motion can be expressed as follows, also known as Navier's equation [211]:

$$(\lambda+\mu)\nabla(\nabla\cdot\mathbf{U})+\mu\nabla^2\mathbf{U}=\rho\frac{\partial^2\mathbf{U}}{\partial t^2}, \tag{1.1}$$

which is equivalent to

$$(\lambda+2\mu)\nabla(\nabla\cdot\mathbf{U})-\mu\nabla^2\mathbf{U}=\rho\frac{\partial^2\mathbf{U}}{\partial t^2}, \tag{1.2}$$

where ∇ and ∇^2 are the gradient operator and the Laplace operator, respectively. λ and μ are the Lamé constants, $\mathbf{U}$ is the vector displacement, and ρ is the density.

The displacement $\mathbf{U}$ can be expressed as a superposition of the gradient of a scalar potential (longitudinal wave components) ϕ and the curl of a vector potential (shear wave components) $\mathbf{\Psi}$ by using the Helmholtz decomposition,

$$\mathbf{U} = \nabla\phi + \nabla\times\mathbf{\Psi}, \quad \text{where} \quad \nabla\cdot\mathbf{\Psi} = 0. \tag{1.3}$$

By substituting Equation (1.3) into (1.1) without considering body forces, Navier's Equation (1.1) can be rewritten as two separate equations: one governing longitudinal waves

$$\frac{\partial^2\phi}{\partial t^2} = c_l^2\nabla^2\phi, \tag{1.4}$$

and the other governing shear waves

$$\frac{\partial^2\mathbf{\Psi}}{\partial t^2} = c_s^2\nabla^2\mathbf{\Psi}, \tag{1.5}$$

where c_l and c_s are the velocities associated with longitudinal waves (L-waves) and shear waves (S-waves) in the infinite isotropic medium, respectively. The relationship between the velocities and the Lam*é* constants can be given by

$$c_l^2 = \frac{\lambda + 2\mu}{\rho}, \tag{1.6}$$

$$c_s^2 = \frac{\mu}{\rho}. \tag{1.7}$$

Two independent equations of (1.4) and (1.5) mean that L-waves and S-waves propagate without interaction in an unbounded isotropic medium. The general solutions to these two equations are given by

$$\phi = \phi_0 e^{i(k_l r - \omega t)}, \tag{1.8}$$

$$\mathbf{\Psi} = \mathbf{\Psi}_0 e^{i(k_s r - \omega t)}, \tag{1.9}$$

where ϕ_0 and $\mathbf{\Psi}_0$ are initial wave amplitude constants. ω is the angular frequency, t is the time, and r is the spatial coordinate of the wave. k_l and k_s represent the wavenumbers of L-waves and S-waves, respectively, and can be expressed as

$$k_{l,s}^2 = \frac{\omega^2}{c_{l,s}^2}. \tag{1.10}$$

Note that ultrasonic bulk waves propagate in infinite media, indicating that boundaries do not affect wave propagation. There are basically two types of bulk waves, namely L-waves c_l and S-waves c_s.

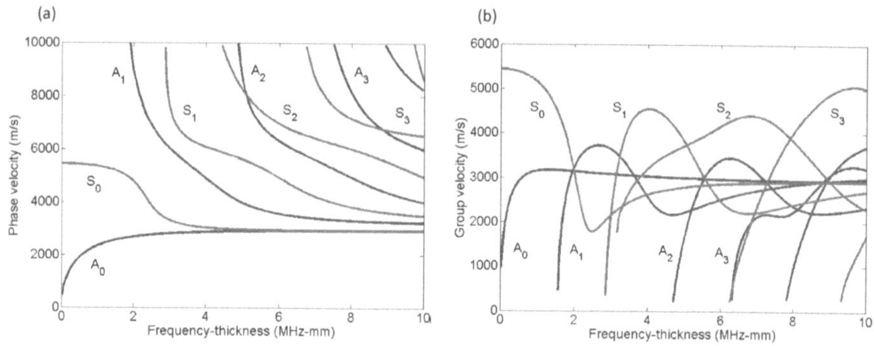

Figure 1.1 Dispersion curves of Lamb modes in an aluminum plate. (a) Phase velocity and (b) group velocity [169].

1.2.2 ULTRASONIC GUIDED WAVES

Ultrasonic guided waves are waves whose energy is concentrated between boundaries and guided along the waveguide. They can be thought of as the superposition of bulk waves reflected back and forth between the boundaries of the structure. In plates, for example, there are two possible guided wave types: Lamb waves and shear horizontal waves. The plate represents the waveguide, and the elastic wave travels parallel to the boundary surfaces. In the direction of the thickness (perpendicular to the bounding surfaces), guided waves behave as standing waves, which determine anti-symmetric, symmetric, and shear horizontal properties [12, 111, 197]. Note that many structures are actually natural waveguides, such as hollow cylinders, rods, and multi-layer structures, if the wavelength is large enough relative to some critical dimension in the waveguide.

In an infinite plate, the Rayleigh-Lamb frequency equations can be obtained by solving Navier's Equations (1.1) for traction-free boundary conditions on both surfaces of the plate. These equations describe the dependence of the wave velocity on frequency, which is referred to as the dispersion curve. The solutions to these equations are obtained numerically, using transfer matrix or global matrix methods [125, 238]. Figure 1.1 shows dispersion curves of anti-symmetric (*A* modes) and symmetric Lamb modes (*S* modes) for an aluminum plate, using a commercially available software package [157]. It can be seen that only fundamental A_0 and S_0 modes exist through all frequencies. The wave modes become complex with increasing frequency-thickness products. When the frequency is fixed, the phase velocity is a function of the thickness, as shown in Figure 1.1(a). These properties of guided waves can be used for thickness estimation. Also, the phase velocity is associated with elastic constants (e.g., Young's modulus) via the dispersion characteristics of the selected guided mode [187], which can be applied to obtain the distribution of elastic constants.

There are several advantages to using ultrasonic guided waves, which are discussed in detail in [197]. We describe these key advantages below.

1. Ultrasonic guided waves enable long-distance inspections from a single probe location, applicable to structures like pipes and rods. There is no need to scan the entire test object under consideration;
2. Generally, even when low-frequency guided wave detection techniques are employed, they can provide higher sensitivity than data obtained from standard localized normal beam ultrasonic testing or other NDT & NDE and SHM techniques;
3. Ultrasonic guided wave techniques allow the detection of hidden structures, coated structures, underwater structures, underground structures, etc. These techniques can become effective because of the combination of single probe position detection using wave structure changes and wave propagation controlled mode sensitivity over long distances.
4. Ultrasonic guided wave inspection is relatively simple and fast, making it cost-effective. For example, in the inspection of long-distance pipelines, besides the locations of the transducers, there is no need to remove the insulation layer on the pipeline or excavate the soil layer covering the pipelines.

1.2.2.1 Guided wave mode selection

To obtain an accurate reconstruction of remnant wall thickness, the selection of suitable Lamb modes for UGWT is very important. In guided waves, many modes can be used to identify damages in plate-like structures. In some frequency-thickness products, the propagation of multiple modes has almost the same velocity. Moreover, in high-frequency-thickness products, many modes exist simultaneously, resulting in complex waveforms. To avoid the overlap of multiple modes, only fundamental modes of A_0 and S_0 below the cut-off frequency-thickness product are considered in most of this book.

In order to choose appropriate fundamental Lamb modes A_0 and S_0 for tomography, it is important to understand its characteristics which affect the performance of UGWT. More details of A_0 and S_0 used in different UGWT applications can be found in [94]. Figure 1.2(a) shows the relationship between the phase velocity and frequency-thickness product for A_0 and S_0 for a free and one-sided liquid-loaded (in this discussion we use water for simplicity) steel plate. It is clear that the fluid loading does not significantly affect the velocities since the curves overlap, suggesting a similar capability of thickness estimation as in a free plate. The most suitable regions for the thickness estimation are the points on dispersion curves where the velocity changes rapidly with the thickness. Such a sensitive region for the A_0 mode exists below around 1.5 MHz-mm (between 0.2 and 1.5 MHz-mm if under water loading), while at higher frequencies the curve becomes flattered and thus loses its sensitivity. In contrast, the S_0 mode has practically no dispersion at low frequencies, and the sensitive region is around 1.3 to 2.1 MHz-mm. The group velocities are shown in Figure 1.2(b). There exists a range in the curves where the S_0 mode is much faster than the A_0 mode, and thus can be easily separated.

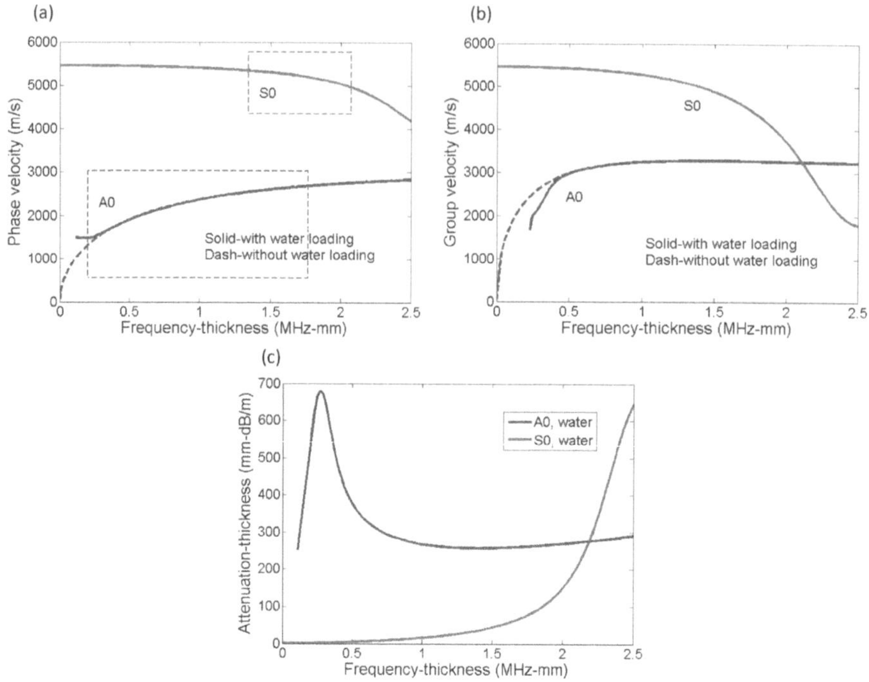

Figure 1.2 The effect of one-sided water-loading on a steel plate. (a) Phase velocity, (b) group velocity, and (c) attenuation dispersion curves for a steel plate without and with liquid loading presenting on one side of the steel plate, represented by solid and dash lines, respectively [169].

The attenuation of fundamental modes is presented in Figure 1.2(c). The vertical axis is given as the product of thickness and attenuation to present the results for the plate of arbitrary thickness. Large attenuation of the A_0 mode prevents its use for long-range UGWT in a liquid-loaded case. The preferred mode is the S_0 mode which has very small attenuation at low frequencies. However, as the frequency increases, the attenuation of S_0 also gradually increases and becomes larger than that of A_0 above 2.2 MHz-mm.

In summary, the advantage of A_0 is the steep gradient of phase velocity in the low frequency-thickness product region, which results in a greater velocity variation with the thickness, thus a better sensitivity. Compared with the A_0 mode, the S_0 mode can propagate faster below 2.2 MHz-mm, making it easier to separate measured signals and avoid reflections from boundaries in some cases. When liquid loading is present, the S_0 mode performs better than the A_0 mode due to the significantly lower attenuation below 1.5 MHz-mm. It needs to be mentioned that these results are also valid for pipes with large diameters for which the curvature of the wall can be neglected [23, 117, 122].

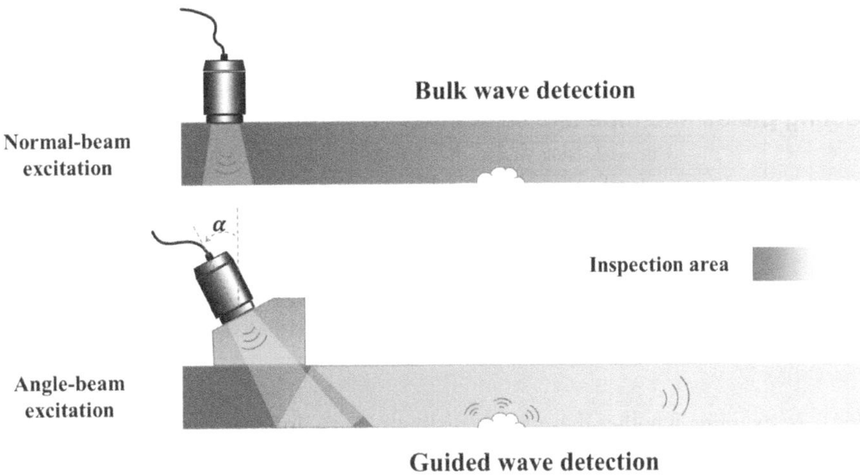

Figure 1.3 Comparison of ultrasonic bulk waves and ultrasonic guided waves.

1.2.3 COMPARISON OF BULK WAVES AND GUIDED WAVES

Figure 1.3 shows a general comparison of ultrasonic inspection using bulk waves and guided waves. Note that in terms of the coverage volume of a structure, ultrasonic bulk waves can only cover a local region below the transducer, whereas ultrasonic guided waves have the ability to cover the total thickness of the structure over long distances. Whether bulk or guided waves are generated is dependent on the frequency used. Lower frequencies corresponding to larger wavelengths can be used in the generation of guided waves. If the guided waves are generated, the wavelength would be larger than the thickness of the structure.

Bulk and guided waves have different characteristics in terms of phase velocities, group velocities, and pulse shape. It is clear that guided waves strongly depend on the frequency and the angle of the impinging wave propagating inside the structure. Guided waves are characterized by dispersive phenomena, while bulk waves are non-dispersive. Phase velocity and group velocity dispersion curves play a very important role in the applications of guided waves in the fields of NDE and SHM. It should be noted that the Rayleigh surface waves in an isotropic elastic half-space are a special case of guided waves, which are non-dispersive.

Further practical comparisons using bulk and guided waves, especially for plate and pipe inspections, are given in Table 1.1.

1.3 NON-DESTRUCTIVE TESTING & EVALUATION AND STRUCTURAL HEALTH MONITORING

The main engineering disciplines related to damage detection or identification include NDT/NDE, SHM, condition monitoring, and statistical process control [72,

Table 1.1
Comparison of ultrasonic bulk waves and guided waves used for detecting plates and pipe-like structures

Ultrasonic Bulk Waves	Ultrasonic Guided Waves
Point-by-point measurement	Global in nature
Time-consuming and tedious	Relatively fast
High-level training	Minimal training
Reflectors must be accessible	Reflectors can be hidden
Fixed distance from the reflector required	Reasonable distance from the reflector acceptable

262]. To examine whether these terms distinguish disciplines, it is useful to have an organizing principle for discussing damage identification issues. It is now generally accepted that five levels are appropriate, as follows [66]: i) detection, ii) localization, iii) classification, iv) assessment, and v) prediction. In detection, a qualitative indication that the damage could be present in the structure can be given. In localization, information on the possible location of the defect can be provided. In classification, information on the damage type is given. In assessment, the estimation of the extent of the damage can be given. In prediction, information on the safety of the structure can be provided.

1.3.1 NON-DESTRUCTIVE TESTING & EVALUATION

NDT, often used interchangeably with NDE, is the use of special equipment and methods to learn about test objects without damaging them. Note that, technically, there are still some differences between NDE and NDT (explain later). Because it allows inspection without affecting the final use of the product, NDT provides an excellent balance between quality control and cost-effectiveness. NDT means that nonliving objects, such as metal, are being evaluated, but the technologies used in NDT are similar to those used in the medical industry. NDT methods are used to ensure that critical components of airplanes, automobiles, and nuclear power plants do not have defects that could result in problems or accidents. The inspection and measurement methods used in NDT are largely based on the scientific principles of physics and chemistry, such as thermography, eddy current, fluorescent penetrant, flux leakage, magnetic particle, radiographic, and ultrasonics.

In contrast to NDT, NDE includes the evaluation of results in addition to testing. NDE is used to describe measurements that are more quantitative in nature. For example, NDE can not only locate defects but also be used to measure information about defects such as shape, size, and orientation. NDE may be used to determine material properties like fracture toughness and other physical characteristics.

1.3.2 STRUCTURAL HEALTH MONITORING

SHM involves observing and analyzing a system using periodically sampled response measurements to monitor changes in the material and geometric properties of engineering structures (e.g., bridges, aircraft, and buildings). SHM is characterized by rapid condition screening, with the goal of providing reliable information about structural integrity in near real-time. The SHM process involves the selection of the excitation method, the sensor types, number, and locations, the data acquisition/storage, data processing, and algorithms. Measurements can be taken to directly detect degradation or damage that may occur to the system, or predict the state of the system by indirectly measuring the size and frequency of loads experienced.

Typically, SHM is associated with online global damage identification in structural systems. NDE is usually performed offline (out of service) in a local manner after the damage has been located. There are exceptions to this rule, as NDE is also used as a monitoring tool for in situ structures such as rails and pressure vessels. NDE is primarily used for characterization and as a severity check when a priori knowledge of the location of the damage is required, such as eddy current and thermography. However, there are exceptions, and a good example is the ultrasonic method. For example, ultrasonic Lamb-wave scattering has the ability to locate damage over a reasonable distance. Guided wave methods have proven to be very powerful in applications such as long-distance pipeline inspection [34]. Lamb waves are singled out where the structure of interest is plate-like, which has proved very useful in the detection of composite laminates [106]. In fact, almost all NDE methods involve taking the structure out of service for more detailed analysis or repair. 'Prediction' is included in the general philosophy of NDE, as code-based detection ultimately links the estimated damage to the estimated consequence based on structural integrity calculations.

Because SHM systems capture time-varying data, environmental and operating conditions can significantly affect the measured data. Therefore, a major challenge for SHM systems is their sensitivity to environmental and operating conditions. Environmental factors include but are not limited to, wind, humidity, moisture, temperature, settlement, seismic actions, and scouring. Operating conditions also include but are not limited to pressure, operating temperature, different loading, and impact. Changing environmental and operating conditions can have a greater impact on structural health than defects-induced changes. Ignoring the influence of these conditions may affect the accuracy of defect detection and may lead to wrong detection results. For example, temperature variations may mainly cause changes in geometry and mechanical properties (e.g., elastic modulus that causes changes in structural stiffness), which can further affect the detection of structural defects.

1.3.3 TEXT PREVIEW

Chapter 2 reviews different ultrasonic tomographic methods such as conventional methods and high resolution methods. In conventional methods, the two typical algorithms of traveltime tomography and diffraction tomography are mainly introduced.

The size of the defect that can be successfully reconstructed is similar to the size of the first Fresnel zone, which is considered approximately the resolution of the travel-time tomography. Resolution can be improved slightly by accounting for refraction in bent-ray tomography. In order to further improve the resolution, it is necessary to consider diffraction. The diffraction tomography, which is limited by the Born approximation, is only valid for low contrast or limited defect size. This chapter also reviews some potential approaches to achieve higher resolution. This part can be divided into two categories: inverse scattering method under the Born approximation and non-linear inverse methods. The strengths and weaknesses of typical approaches are discussed here.

Chapter 3 outlines the Lamb wave diffraction tomography based on Mindlin's plate theory. The formulation of Mindlin's plate theory is first introduced, after which damage modeling, the scattering integral equation, the Born approximation, and the multi-static data matrix are presented. Although the analytical Green's function is used here, numerically computed Green's functions can also be applied to consider structural complexities. Three-dimensional FE simulations are used to reconstruct damaged images for circular, cross-shaped, and square-shaped thickness reductions. In all cases, the thickness reduction is 10% of the plate thickness, and the reconstructed images accurately predict both the thickness reduction and the damage geometry.

Chapter 4 introduces the theory of UGWT based on the frequency-domain mono-parameter full waveform inversion (FWI), which consists of the forward modeling and the inversion method. The thickness reconstructions of a single circular hole and an irregularly shaped defect from both simulations and experiments are investigated. The UGWT based on the frequency-domain mono-parameter FWI can achieve improved resolution, however, the limitations are also discussed.

In Chapter 5, the resolution and accuracy analysis of UGWT based on the frequency-domain mono-parameter FWI is presented. Simulation studies of smoothly varying defects with different widths and depths over a range of frequencies confirm the achievable resolution based on both the acoustic and elastic models. Furthermore, thickness reconstructions of multiple defects are validated by the experiment and the results are compared with diffraction tomography results.

Chapters 4 and 5 are concerned with the full viewing problem, but in practical applications, limited access to waveguides containing corrosion defects restricts the number and range of viewing angles. An adaptive threshold regularization method to predict missing components, combined with the FWI algorithm for limited view configurations, is used to improve image quality in Chapter 6.

In Chapter 7, in the application of corrosion mapping, UGWT of liquid-loaded plates based on the frequency-domain mono-parameter FWI is presented. This part with the consideration of attenuation for reconstructing the thickness is investigated using different defect sizes and frequencies. Then, a newly developed UGWT system using self-designed piezoelectric transducers for corrosion monitoring of plate-like and pipe structures is introduced. The accuracy and efficiency of this new system are demonstrated by continuously forced corrosion experiments. Note that the thickness mapping of liquid-loaded pipes with two ring transducer arrays requires consideration of limited view problems and attenuation.

On the basis of the above, the UGWT based on frequency-domain mono-parameter FWI is extended to reconstruct Young's modulus and consider anisotropy in laminated composites in Chapter 8. First, the principle of reconstructing Young's modulus from the velocity is presented and the effectiveness of this method is verified by simulations, and an experiment with different defect types and sizes. Second, previous studies focus on isotropic materials, and here we further concentrate on anisotropic materials and improve the imaging accuracy by considering anisotropy in UGWT based on FWI.

In Chapter 9, the improved mono-parameter FWI in UBWT uses different descent methods to solve the optimization problem. The proposed gradient sampling FWI (GSFWI) combined with reverse time migration for reconstructing defects hidden in the multiple material parts manufactured by 3D printing. The low frequency GSFWI is used to obtain the interface of the unknown background, and then this result is used as the starting model for the high frequency reverse time migration to achieve the reflectivity image of defects in 3D printed parts. The performance of this method is analyzed using experimental data and then cross-checks with X-ray CT.

Finally, in the quality control application presented in Chapter 10, the multi-parameter FWI in UBWT is developed for simultaneous reconstructions of the velocity and density of metal parts. Compared with mono-parameter inversion, multi-parameter inversion increases indetermination. Therefore, the truncated Gauss-Newton method, which is used to estimate the inverse Hessian, is used to mitigate the effects of cross-talk between velocity and density. This algorithm is validated by both the simulation and experiment.

1.3.4 SUMMARY

A brief overview of ultrasonic bulk waves, which can propagate in two types of longitudinal waves and shear waves that are uncoupled in an infinite medium, and ultrasonic guided waves that require a structural boundary for propagation, is presented. Ultrasonic guided wave tomography (UGWT) uses guided waves, which can be considered as a superposition of bulk waves reflected back and forth between the boundaries of the structure, along with tomographic methods. UGWT provides a solution to quantify thickness losses by corrosion. The dispersion characteristics of selected guided modes can be exploited to achieve thickness reconstructions to be generated from velocity reconstructions based on acoustic velocity inversion algorithms. Additionally, due to the presence of multiple modes, it is necessary to select suitable modes for UGWT. UGWT has also been extended to reconstruct material properties. Compared to UGWT, UBWT is often less complicated and easier to handle, making it suitable for detecting deeper defects in thicker structures.

Then, two other important differences between SHM and NDT & NDE are emphasized. The first difference lies in the sensing technology. SHM typically relies on permanently installed, fixed-position sensors, while NDT & NDE instruments, such as thermographic cameras, are usually brought to the structure of interest on an as-needed basis. The other difference is that many SHM experts consider that SHM should be carried out online or in an automated manner, whereas NDT & NDE may be performed occasionally or only when necessitated by other inspections. The transition from NDT & NDE to SHM technology hinges on the affordability of sensing instruments, making them economical for permanent installation with automatic triggering and data acquisition. In fact, with advancements in technology, durable sensors can now be permanently installed to obtain valid and reliable measurements.

Ultrasonic inspection has transitioned from the NDT & NDE technology to the SHM technology, in part due to the development of inexpensive sensors (such as piezoelectric patches) and the utilization of guided waves. These advancements mean that ultrasonic waves can travel large distances; for example, Lamb waves in plates significantly reduce the need for dense sensor placement [49]. The use of phased arrays is another advance in ultrasonic NDT & NDE. A phased array is an ultrasonic transducer with multiple elements which can be triggered to actuate in a prescribed sequence. Phased arrays are useful for inspection because the waves can be steered and focused without moving the probe. Some designs allow the sensor to be bonded to the structure, essentially creating an SHM system, and a good example of this can be found in [10].

2 Ultrasonic Tomography Methods

2.1 INTRODUCTION

Ultrasonic inspection of engineering structures has been widely used in the aviation, automotive, and petrochemical industries. Conventional inspection techniques are largely used to determine the locations and sizes of defects and inclusions, without providing material properties. With the increasing use of high-performance engineering components, there is a substantial need for more detailed information on material properties to assess the failure probability of structures in these applications. Ultrasonic tomography emerges as a promising technique for obtaining quantitative information on material properties. In ultrasonic tomography, there are two major steps to be considered: First, quantitative information is extracted from ultrasonic waveforms using various techniques. Second, the obtained information must be appropriately processed to produce a mapping of the object parameter using tomographic methods [62]. This chapter focuses on a literature review of ultrasonic tomographic methods for NDT & NDE and SHM.

2.2 CONVENTIONAL METHODS

Tomography is the process of imaging by sections or slices using a penetrating wave. Common methods involve radiology or X-rays, lasers, and ultrasound. In most cases, images are produced using a mathematical procedure called tomographic reconstruction. For example, X-ray computed tomography (CT) is technically produced from multiple projectional radiographs. The aim of CT is to obtain information about the nature of the materials located at exact positions inside the test object. There are many different reconstruction algorithms. These algorithms provide a finite sequence of explicit instructions that guide the step-by-step process from a given input to the desired output. These reconstruction algorithms are based on the assumption that the slice is infinitely thin. Effective algorithms are critically important. Despite the limited and sometimes inaccurate nature of the measured data, there have been considerable efforts to develop algorithms that are both fast and capable of producing acceptable reconstructions when implemented on a computer [146].

2.2.1 TRAVELTIME TOMOGRAPHY

Ultrasonic traveltime tomography (TT) is a method used for estimating the ultrasonic velocity structure utilizing a given set of arrival times. The straight ray approximation represents the simplest case. The straight ray TT, originating from X-ray CT,

DOI: 10.1201/9781003667926-2

operates under the assumption that ray paths are straight lines connecting transmitters and receivers, where refraction and diffraction effects of waves are disregarded. The straight ray TT uses the arrival times of wave-packets to reconstruct the distributions of slowness (the reciprocal of velocity). As such, accurately extracting the arrival times from the time traces is a crucial step in ray model-based reconstruction. Various approaches for this have been developed, including cross-correlation, thresholding, and wavelet transforms [61, 163, 271]. Tomographic methods are then employed to generate the velocity map, using techniques based on direct inversion (such as the filtered back-projection method [103]) and iterative gradient-based inversion [84]. Besides the straight ray TT, the bent ray TT has also been developed to take refractive effects into account, but not diffraction effects [120]. As the tomographic problem becomes nonlinear, conjugate gradient methods are usually implemented [90, 121].

The resolution limit of the straight ray TT is dependent on the accuracy of the arrival time extraction. When the cross-correlation method is used to pick arrival times, the resolution limit is approximately the size of the first Fresnel zone $\sqrt{L\lambda}$, where λ is the wavelength of the illuminating wavefield and L is the distance between the source and the receiver [260]. In other words, the size of the test object which can be reconstructed well is similar to the first Fresnel zone. The resolution can be improved when arrival times can be estimated from the signal first break [259]; however, in practice, the accuracy of first break arrival measurements is affected by noise. It is possible to slightly increase the resolution of the bent ray TT because refraction is considered, but further increases in resolution are not achievable without accounting for diffraction. In general, ignoring refraction limits the algorithms to the detection of low-contrast defects, and ignoring diffraction limits the algorithms to large, smoothly varying test objects [97].

2.2.2 DIFFRACTION TOMOGRAPHY

Diffraction Tomography (DT) was first proposed by Malyarenko [130] and Rohde [196]. This method incorporates diffraction and scattering effects into tomographic algorithms to enhance both the resolution and quality of reconstructions. DT is founded on a linearized scattering model, which includes the Born approximation or the Rytov approximation [22]. However, for the purpose of this discussion, only the Born approximation is considered here.

The Born approximation assumes that the scatterer is small and exhibits low contrast, in other words, it is a weakly scattering object. This means that the phase change between the incident field and the field inside the scatterer is minimal. Under this approximation, the field scattered by the scatterer can be mapped to the spatial frequency domain Ω, as depicted in Figure 2.1. An array illuminates a scatterer from direction $\bar{r_0}$ and the scattered field is measured in direction $\bar{r}$ (Figure 2.1(a)). This is then mapped to the point $k_0(\bar{r_0} - \bar{r})$ in the Ω domain (k_0 is the wavenumber), as shown in Figure 2.1(b). As the transmitter location rotates around this circular array, a circle in the Ω domain with radius $2k_0$ is obtained, termed the Ewald limiting circle [65]. This means that the highest spatial frequency component is $2k_0$, i.e., the

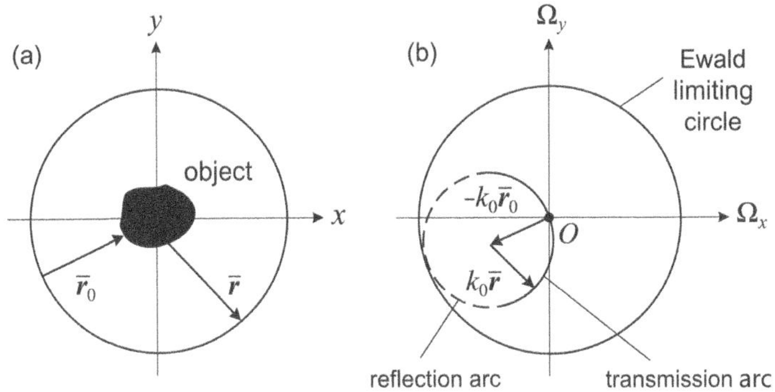

Figure 2.1 Illumination of the object and the subsequent scattered field. (a) The object is illuminated from direction $\bar{r_0}$, and the scattered field is measured from direction $\bar{r}$. (b) Under the Born approximation, the scattered field measured in the $\bar{r}$ direction, originating from illumination in the $\bar{r_0}$ direction, is mapped to point $\Omega = k_0(\bar{r_0} - \bar{r})$ in the spatial frequency domain [169].

resolution is $\lambda/2$. However, when limited access to the scatterer restricts the range of viewing angles, the coverage region is no longer the entire circle, leading to a resolution reduction. In transmission mode DT, the scattered field only contains information from the transmission zone, which maps in the Ω domain to the transmission arc. Consequently, the maximum reconstruction resolution reduces to $\sqrt{2}k_0$ [196]. In reflection mode DT, the reflection zone maps to the reflection arc in the Ω domain, and the resolution varies from $\sqrt{2}k_0$ to $2k_0$, with only high frequencies available in this coverage region [7]. DT only works effectively with small or low-contrast defects, where the phase shift traveling through the defect is minimal.

Researchers have proposed using the straight ray TT to provide initial velocity estimates for wave equation-based DT. This approach leverages the lower resolution TT, which is insensitive to the starting model, and then utilizes DT's ability to improve resolution based on accurate input velocity information. More recently, Huthwaite et al. [96] combined the bent ray TT and DT algorithms to propose the HARBUT algorithm. Initially developed for medical applications, it was later extended to NDT & NDE applications [97]. The HARBUT algorithm uses BRT for background estimation to correct the Born approximation used in the DT algorithm. This method expands the applicability range of DT reconstruction, as it can estimate and account for phase distortion through defects. However, the reconstruction of small and high-contrast defects still faces degradation. To address this, iterative HARBUT is used to improve reconstruction quality [97]. This process uses the result of the last HARBUT iteration as the starting model for the next stage, repeating this procedure until a convergence criterion is reached. This hybrid algorithm provides a fast, high-resolution reconstruction of ultrasonic velocity, with the same theoretical resolution limit as the DT algorithm, which is $\lambda/2$.

2.3 HIGH RESOLUTION METHODS

The primary parameter of ultrasonic wave imaging is resolution, which determines the system's ability to discern the smallest details in an object. The resolution of most imaging techniques is constrained by the diffraction limit related to the wavelength that can be propagated [78, 214]. This inherently limits the possibility of achieving sub-wavelength resolution. Therefore, to enhance resolution, an intuitive choice is to reduce the wavelength (i.e., increase the operating frequency). However, as the wavelength decreases, wave attenuation increases, leading to a reduction in the maximum imaging depth [214]. Using high-power equipment to amplify excitation energy, and thus the imaging range comes at a high cost. Moreover, as the operating frequency increases, the upper frequency bound is limited by coherent noise and the emergence of multiple scattering. It is important to note that in most practical applications, the achievable resolution limit does not reach the diffraction limit due to various constraining factors (e.g., the increase in beam width and the large grain size in cast structures). In this section, some potential imaging approaches for achieving higher resolution are presented, such as the inverse scattering approach under the Born approximation and non-linear inverse methods.

In principle, most ultrasonic imaging methods capable of achieving high or super-resolution rely on evanescent waves or multiple scattering [109]. Many other approaches to super-resolution involve near-field imaging using high-frequency evanescent waves generated by the interaction of low-frequency incident waves with small defects. Techniques using evanescent waves demonstrate that the sub-wavelength resolution achieved does not contradict the diffraction limit. However, these evanescent waves do not propagate and decay exponentially at approximately one wavelength distance. As a result, methods based on evanescent waves are only suitable for defects located near the surface. Some studies have reported using meta-materials to amplify evanescent waves in order to achieve sub-wavelength resolution.

For instance, Amireddy et al. [5] experimentally demonstrated sub-wavelength ultrasound imaging of defects in metal using holey-structured meta-material lenses. Their results highlighted the extraordinary transmission capability of this meta-material, achieved by coupling higher frequencies in the incident ultrasonic field to the resonant modes of the lens. The key to this success involves optimizing the geometric parameters of the holey lenses to provide suitable high-frequency resonant filters. This allows for the detection of evanescent waves with a high SNR. Laureti et al. [113] utilized entrapped air meta-materials, manufactured via vat photopolymerization, for ultrasonic super-resolution imaging in water at frequencies between 200-300 kHz. Their findings demonstrated that this approach outperforms conventional designs of polymer-based meta-materials, providing a promising direction for the creation of polymer-based ultrasonic meta-materials in water using additive manufacturing.

As previously mentioned, a fundamental challenge lies in achieving sub-wavelength resolution using evanescent waves, as it is only applicable in the near-field zone—meaning at least one of the probing transducers should be within a distance of 1λ from the defect. The progress made in the near-field zone naturally raises the question of whether super-resolution can also be achieved when all probing transducers are placed in the far field. Overcoming this challenge is crucial to break free from the wavelength constraint. For instance, Simonetti [215, 216] demonstrated that sub-wavelength resolution in the far-field zone could be achieved by incorporating multiple scattering in the imaging algorithm. This suggests that the distortion of the wavefield caused by multiple scattering within the test object encodes sub-wavelength information into the far-field pattern of the scattered waves.

Additional research has focused on overcoming the diffraction limit to achieve super-resolution through computational methods. Song et al. [225] proposed a deep learning-enhanced super-resolution ultrasonic beamforming approach. This approach initially creates focused images using a total focusing method based on full matrix capture, and then applies a fully convolutional network (FCN) architecture to locally resolve sub-wavelength scale fine details of defects. Rao et al. [179] proposed an ultrasonic method based on a supervised FCN, designed to quantitatively reconstruct high contrast, small defects ($< \lambda/2$) in multi-layered bonded composites. This method includes both training and prediction processes. In the training process, an FCN establishes a non-linear mapping from measured ultrasonic data to the corresponding L-wave velocity models of multi-layered bonded composites. In the prediction process, the network obtained from the training is used to directly reconstruct the L-wave velocity models from newly measured ultrasonic data of adhesively bonded composites.

In the remainder of this chapter, the discussion is based on the far-field hypothesis (greater than λ). The resolution limit can be traced back to the loss of information related to the object's sub-wavelength spatial periodicity, which is encoded in the evanescent component of fields scattered by the test object, when these fields are monitored from a distance far from the object [261].

2.3.1 INVERSE SCATTERING METHOD UNDER THE BORN APPROXIMATION

In inverse scattering problems, while the incident field is known and the measured scattered field is obtained, the objective function remains unknown. Hence, the problem is to determine the objective function based on the measured scattered field. The reconstruction of the objective function can produce an image of object properties, hence this problem essentially involves creating an image of the object from the scattered field [67].

The inverse scattering problem is inherently ill-posed. Under the Born approximation, the forward scattering problem is simplified by ignoring the effects of multiple scattering, rendering this problem linear, with the inverse problem also being a linear one relating to the object function. This approximation establishes a one-to-one relationship between the scattering amplitude and the spatial Fourier transform of the object function. However, this relationship is confined to the Ewald limiting circle of radius $2k_0$ in the Ω domain, as depicted in Figure 2.1. Consequently, the determination of the object function is only based on the information within this circle. The Born approximation does not account for the information outside this circle that is contained in the object function. One potential approach to handle this limitation is to set the information outside this circle to zero, as follows:

$$\tilde{O}_{ba}(\Omega) = H(\Omega)\tilde{O}(\Omega), \tag{2.1}$$

where $\tilde{O}_{ba}(\Omega)$ is the Born approximation of the object function $\tilde{O}(\Omega)$ in Ω domain. $H(\Omega)$ presents a low-pass filter, $H(\Omega) = 1$ for $|\Omega| \leq 2k_0$ and $H(\Omega) = 0$ for $|\Omega| > 2k_0$. This implies that spatial frequencies larger than $2k_0$ are not recovered, thereby establishing the classical resolution limit at $\lambda/2$.

Equation (2.1) suggests that $\tilde{O}_{ba}(\Omega)$ is defined over the entire spatial frequency domain, Ω. The reconstructed image of $O_{ba}(r)$ in geometric space can be achieved by taking the inverse Fourier transform of $\tilde{O}_{ba}(\Omega)$. This is given by the convolution of the point spread function $h(r)$ and the actual object function $O(r)$ [216],

$$O_{ba}(r) = \int_D d^2r' h(r-r')O(r'), \tag{2.2}$$

where

$$h(r) = k_0 \frac{J_1(2k_0|r|)}{\pi|r|}. \tag{2.3}$$

$h(r)$ is the inverse Fourier transform of low-pass filter $H(\Omega)$, and J_1 is the Bessel function of the first order. $O_{ba}(r)$ is a low-pass filter version of $O(r)$ without spatial frequencies larger than $2k_0$.

The Born approximation allows for the potential solution of obtaining super-resolution based on the analytic function $\tilde{O}(\Omega)$ (see Equation (2.1)). This function, $\tilde{O}(\Omega)$, can be extrapolated to points outside the Ewald limiting disc by utilizing analytic continuation [1]. By taking the inverse Fourier Transform of the data, the exact object function can be obtained instead of the low-pass filtered version [67]. This method is akin to deconvolution of $O(r)$ in Equation (2.2), using information about the point spread function. Hence, this approach could theoretically achieve unlimited resolution. However, several authors [201, 222, 240] have demonstrated that analytic continuation is highly unstable and sensitive to noise. Consequently, it cannot provide stable solutions to inverse scattering problems.

The second approach for achieving super-resolution under the Born approximation involves incorporating information regarding object properties, such as structure or material characteristics, along with regularized data. However, this method is not always feasible since such information is often not known in advance.

2.3.2 NON-LINEAR INVERSE METHODS

Reconstruction of the object function, taking into account the effects of multiple scattering, constitutes an ill-posed non-linear problem. Two main strategies for tackling this non-linear inverse problem are presented, namely Linear Sampling Methods (LSM) and iterative methods [127, 161].

2.3.2.1 Linear sampling method

The LSM, proposed by Colton et al. [47], offers a means of replacing a non-linear inverse problem with a linear integral problem. LSM facilitates the reconstruction of the object shape from the far-field T-matrix operator T_∞: $L^2(s) \to L^2(s)$,

$$T_\infty|y\rangle = \int_s ds(\bar{r_0}) f(k_0\bar{r}, k_0\bar{r_0}) y(k_0\bar{r_0}), \tag{2.4}$$

where L^2 denotes the Hilbert space; s represents the unit shell, and the ket symbol is used based on Dirac notation [1]. $f(k_0\bar{r}, k_0\bar{r_0})$ is the scattering amplitude, which is obtained by illuminating the object from the incident directions r_0 and detecting the scattered field in the r directions (see Figure 2.1). Importantly, the scattering amplitude in LSM includes the effects of multiple scattering. $y(k_0\bar{r_0})$ denotes the amplitude of the incident wave.

LSM considers the solution to this equation

$$T_\infty|y\rangle = |g_\infty(\bar{r}, z)\rangle, \tag{2.5}$$

where $g_\infty(\bar{r}, z) = e^{-ik_0\bar{r}\cdot z}$ is the steering function and proportional to the far field pattern of the source at a point z in $\bar{r}$ direction. The $L^2(s)$ norm of the solution $\sqrt{\langle y_z|y_z\rangle}$ becomes unbounded as the point z approaches the boundary ∂D of the support D (i.e., the volume occupied by the object) of the object function [216]. Thus, ∂D can be determined by estimating $\sqrt{\langle y_z|y_z\rangle}$, which is expressed as the locus of points z where $\sqrt{\langle y_z|y_z\rangle}$ exhibits significant increases.

The Factorization Method (FM) [108], proposed later, is a modified version of LSM. This method considers the following equation:

$$(T_\infty^\dagger T_\infty)^{1/4}|y\rangle = |g_\infty(\bar{r}, z)\rangle, \tag{2.6}$$

where $T_\infty^\dagger$ is the adjoint of T_∞ and $T_\infty^\dagger T_\infty$ represents the time reversal operator [162]. If Equation (2.6) is solvable, there exists a z that belongs to the support D. Based on Picard's theorem [48], Equation (2.6) is solvable if and only if

$$\sum_{j=1}^{\infty} \frac{1}{|\mu_j|} |\langle g_\infty(\bar{r}, z)|v_j\rangle|^2 < \infty, \tag{2.7}$$

where μ_j and v_j are eigenvalues and eigenfunctions of T_∞. Similarly, as in LSM, the support D is the locus of points z for which Equation (2.7) is finite,

$$\mathrm{FM}(z) = \left(\sum_{j=1}^{\infty} \frac{1}{|\mu_j|} |\langle g_\infty(\bar{r}, z)|v_j\rangle|^2 \right)^{-1} > 0. \tag{2.8}$$

FM applies Equation (2.8) to each coordinate within the area of interest. If a coordinate is part of the support of the object D, this equation will be non-zero; otherwise, it will be zero. Therefore, D will be presented by points with non-zero values. According to Equation (2.8), in theory, FM can achieve unlimited resolution

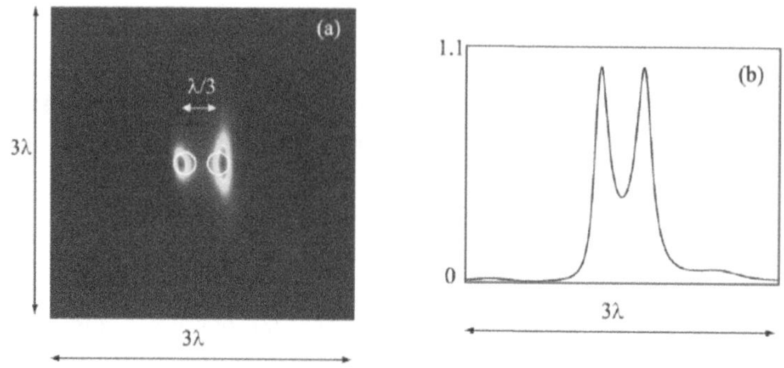

Figure 2.2 Super resolved images at 42 kHz around two scatterers. (a) FM result, and (b) a cross-section of the FM pseudo spectrum along the horizontal center line [216].

in the reconstruction of the object shape. In the presence of noise, FM has been shown to obtain super-resolution of $\lambda/3$ in the real experiment [216], as shown in Figure 2.2.

In both LSM and FM, no assumptions are made regarding the Born approximation, thus taking into account multiple scattering within the object, which leads to super-resolution [216]. However, LSM and FM cannot be used to reconstruct the material properties of the object. Furthermore, they are sensitive to noise and experimental uncertainties.

2.3.2.2 Iterative methods

Iterative algorithms can also be used to solve non-linear inverse scattering problems. In these algorithms, measured scattered data is utilized to construct approximate solutions. The forward problem is then applied to derive the scattered field from these approximations, which is subsequently compared with the measured scattered field. The results of this comparison are used to update the approximation. This iterative process is repeated until the modeled data converges with the measured data.

Distorted Born iterative method

The Distorted Born Iterative Method (DBIM) solves the non-linear inverse scattering problem by calculating iterative solutions using successive linear approximations [39, 83]. This approach takes into account the effects of multiple scattering. To gain a better understanding of DBIM, it is also important to present the primary principle of the Born Iterative Method (BIM). More details about DBIM and BIM can be found in [39, 257].

When the incident field (pressure) is denoted as $U_{inc}(r)$, the scattered field $U_{sca}(r)$ can be written as the integral equation [126]

$$U_{sca}(r) = \int_S dr' G_{reen0}(k_{bac}, r - r') O(r') U(r'), \tag{2.9}$$

where the total field $U(r)$ is equal to $U_{inc}(r) + U_{sca}(r)$ in the inhomogeneous area S and $G_{reen0}(k_{bac}, r - r')$ represent Green's function for the background. k_{bac} is the wave number, and $O(r)$ represents the object function. Equation (2.9) can also be given by

$$U(r) = U_{inc}(r) + \int_S dr' G_{reen0}(k_{bac}, r - r') O(r') U(r'). \tag{2.10}$$

Equation (2.10) is a non-linear integral expression of the object function $O(r)$. The inverse problem consists of solving for $O(r)$ based on the known scattered field.

The basic principle of BIM is presented first. The unknown $O(r)$ is initially assumed to be zero, which corresponds to $U(r) = U_{inc}(r)$. Then, the total field $U(r)$ in the integrand in Equation (2.10) is replaced by $U_{inc}(r)$, and $O(r) \simeq O_0(r)$ is solved. Next, $O(r)$ is substituted by $O_0(r)$ in the integrand, and $U(r)$ is obtained. $O(r) \simeq O_1(r)$ is solved based on the obtained total field $U(r)$. This process is repeated until the convergence criterion is met. In all iterations, Green's function G_{reen0} remains unchanged in BIM.

In contrast, DBIM introduces a difference $G_{reen0}(k_{bac}, r - r')$, which is renewed to $G_{reen}(k_1, r - r')$ in the first iteration, then to $G_{reen}(k_2, r - r')$ in the second iteration, and so on [126]. The k_1 and k_2 correspond to the object functions O_1 and O_2, respectively. It should be noted that the total field is updated in each run based on the most recently obtained object function. In summary, while BIM keeps the background fixed and only updates the total field, DBIM does not restrict the background to homogeneity and updates it in each iteration. Consequently, both the total field and the Green's function are updated.

DBIM can achieve super-resolution, attributed to the effects of multiple scattering within an inhomogeneous object [39]. The object's high-resolution information is carried by evanescent waves under single scattering physics. In the multiply scattered field, a conversion occurs between evanescent waves and propagating waves. Therefore, the measured scattered waves contain high-resolution information about the object due to this conversion. DBIM can extract this information, taking multiple scattering into account. The super-resolution capability of DBIM has been validated experimentally [39], as shown in Figure 2.3. However, this method is computationally expensive because it requires repeated computation of solutions for each necessary forward scattering problem.

Markov-Chain Monte Carlo algorithm

A Bayesian method has been employed to reconstruct sub-wavelength defects from their scattered fields in elastic waveguides [144]. The main idea of this method is to match the modeled scattered field with the measured scattered field using a Markov-Chain Monte Carlo (MCMC) algorithm [139], in conjunction with efficient models. The model's output data is obtained using an analytical model [143] and/or an efficient finite element model [245], which are used to estimate the full 3D geometry of sub-wavelength defects in elastic waveguides. Subsequently, statistical information about the defect properties (e.g., shape, depth, and size) can be acquired. Compared to velocity inversion approaches, this algorithm can capture guided wave

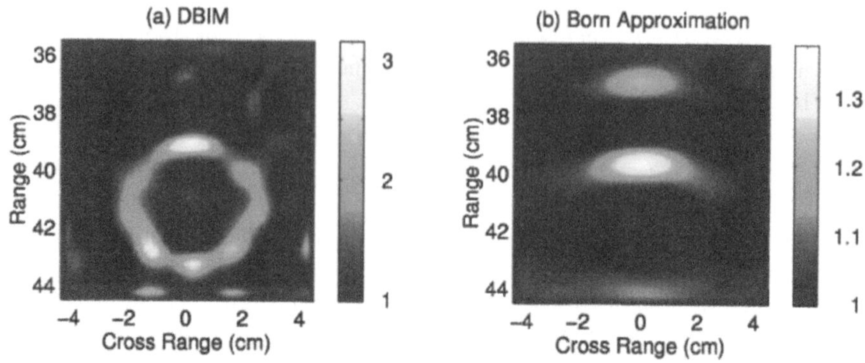

Figure 2.3 (a) Permittivity reconstruction of a plastic pipe with an outer diameter of 4.8 cm and the inner diameter of 3.7 cm by using distorted Born iterative method (DBIM). (b) Reconstruction of this plastic pipe under the Born approximation [39].

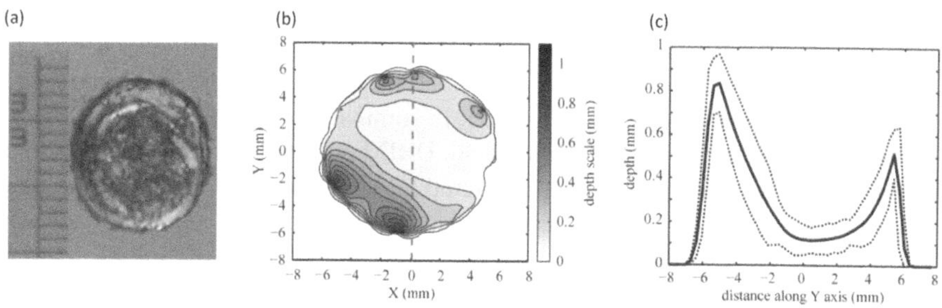

Figure 2.4 (a) Photography of a real defect, (b) estimated defect profile from the finite element model, and (c) cross-section of (b) along X = 0 mm (solid line) and the standard deviation (dotted lines) [144].

scattering rather than acoustic scattering, enabling higher resolution. The MCMC algorithm has been shown to accurately reconstruct various sub-wavelength defects through both simulations and experiments. Furthermore, accurate reconstructions can be obtained even with partial data on the scattered field, as shown in Figure 2.4. However, this MCMC algorithm is inefficient as it typically requires thousands of iterations and is also dependent on the efficiency of the forward model.

Full waveform inversion

Full waveform inversion (FWI) is an iterative optimization approach that yields high-resolution models [165, 235, 249]. FWI estimates model parameters by fitting modeled data to the measured (observed) data, where the modeled data are computed by solving a full-wave equation. Unlike TT, which relies on ray assumptions, FWI takes into account higher-order diffraction and scattering. This approach can improve resolution compared to DT, which is limited by linear scattering.

In the FWI method, the forward model can be performed in either the time or frequency domain. Time domain FWI covers the entire frequency spectrum, providing a precise solution that can be easily parallelized [142, 208, 235]. Temporal windowing in the time domain FWI can be used to reduce computation. On the other hand, frequency domain FWI facilitates easier handling of multiple sources, as the forward modeling operator can be reused in each iteration [132, 164, 165]. Additionally, a multi-scale strategy in frequency domain FWI can mitigate the nonlinearity of the inverse problem by gradually progressing from low to high frequencies, increasing the likelihood of obtaining a globally optimal solution [220]. Finally, dispersion and attenuation can be seamlessly incorporated through complex velocity, without incurring additional computational cost in the formulation [163].

2.4 SUMMARY

The literature review summarizes two main tomographic methods commonly used in geophysical and medical imaging: Traveltime Tomography (TT) and Diffraction Tomography (DT). TT, which is based on the ray theory assumption, is suitable for imaging objects larger than the first Fresnel zone. It utilizes the travel time of waves propagating through the object to reconstruct subsurface or internal structures. Bent ray TT is an extension of TT that takes into account refraction effects, allowing for a slight improvement in resolution compared to traditional TT. By considering the bending of rays due to velocity variations, Bent ray TT can provide more accurate imaging results for complex subsurface structures.

DT, on the other hand, incorporates diffraction effects into the imaging process, resulting in an improved resolution limit compared to TT. It considers the scattering of waves by the object and utilizes the Born approximation to reconstruct object properties. However, DT is only valid for small or low-contrast defects. By combining the merits of bent ray TT and DT algorithms in the Hybrid Algorithm for Refraction and Diffraction-Based Ultrasonic Tomography (HARBUT), the range of applicability of DT can be greatly increased. HARBUT enhances the resolution of ultrasonic tomography by incorporating both refraction and diffraction effects, providing more accurate imaging results for a wider range of objects.

To achieve higher resolution in imaging, several potential approaches are reviewed, including inverse scattering methods under the Born approximation and non-linear inverse methods. Under the Born approximation, the inverse scattering problems can be linearized by neglecting multiple scattering effects. Analytic continuation is one approach that can be used to extrapolate the object function beyond the Ewald limiting circle, theoretically achieving unlimited resolution. However, analytic continuation is not stable and is highly sensitive to noise.

To address the non-linear inverse problem, two methods are discussed: linear sampling methods (LSM) and iterative methods. The Factorization Method (FM), a modified version of LSM, is described, which considers multiple scattering effects and can achieve super-resolution images of object shapes. However, FM does not provide information about the material properties of the object and is sensitive to noise.

Three iterative approaches are then discussed: the distorted Born iterative method (DBIM), the Markov-Chain Monte Carlo algorithm (MCMC), and full waveform inversion (FWI). DBIM has demonstrated super-resolution in experiments but is computationally expensive. MCMC in UGWT captures guided wave scattering behavior for improved resolution but requires a large number of iterations for convergence. FWI overcomes limitations associated with ignoring low-frequency effects in TT and allows for higher-order diffraction and scattering effects in its numerical solver. This enables the capture of more accurate wave scattering physics and potentially leads to more precise inversion results. The FWI algorithm will be investigated further in this book. However, before diving into FWI, the next chapter will introduce DT for comparison with FWI.

2.5 EXERCISES

1. What are the types of tomography?
2. What are the advantages and disadvantages of traveltime tomography?
3. What are the advantages and disadvantages of diffraction tomography?
4. How to choose whether to use traveltime tomography or the diffraction tomography in the application?
5. What are the methods to achieve high resolution?
6. What is the diffraction limit? What is the reason for the diffraction limit?
7. What are the characteristics of the linear sampling method? What are the strengths and weaknesses of this approach?
8. What methods are included in the iterative method? What are the characteristics of each method?
9. What is the full waveform inversion? What are the main features of this method? What are the advantages and disadvantages?
10. What other methods do you think can achieve super-resolution imaging? What is the main reason?

3 Mindlin Plate Theory and Lamb Wave Diffraction Tomography

3.1 INTRODUCTION

Mindlin plate theory extends classical Kirchhoff plate theory by incorporating the effects of transverse shear deformation and rotary inertia. While Kirchhoff's theory is primarily applicable to thin plates where shear effects are negligible, Mindlin plate theory provides more accurate predictions for thick plates and laminated composite structures, where transverse shear cannot be ignored. By accounting for shear strains and relaxing the assumption that normal sections remaining perpendicular to the mid-surface, Mindlin's formulation captures the essential physics needed to describe higher-order vibrational modes and complex wave behavior in real-world structural components. This theory forms a foundational basis for modeling Lamb waves and interpreting their interactions with defects in structures of finite thickness.

Lamb wave diffraction tomography is a powerful imaging technique that reconstructs the internal features of a structure by analyzing the scattering and diffraction of guided waves. In this approach, Lamb waves—elastic guided waves that propagate in plates—are transmitted and received across a network of sensors or scanning points. The scattered fields generated by internal anomalies such as cracks, delaminations, or inclusions are collected and processed inversion algorithms based on diffraction tomography principles. By treating the scattered waves as projections of the internal discontinuities, diffraction tomography synthesizes high-resolution images of material defects, providing both qualitative and quantitative insights. Combining the dispersive characteristics of Lamb waves with diffraction-based imaging enables detailed characterization of complex damage geometries in engineering structures, thus supporting early-stage damage detection and integrity assessment.

The integration of Mindlin plate theory and Lamb wave diffraction tomography provides a rigorous and efficient framework for simulating, interpreting, and imaging guided wave interactions in real-world plate-like structures.

3.2 MINDLIN PLATE THEORY FOR LAMB WAVE

In Mindlin plate theory [141], the flexural wave on thin plates can be expressed in the Cartesian coordinate system using the following displacement components (u_x,

DOI: 10.1201/9781003667926-3

u_y, u_z) as

$$u_x = -z\psi_x(x,y,t), \tag{3.1}$$
$$u_y = -z\psi_y(x,y,t), \tag{3.2}$$
$$u_z = w(x,y,t), \tag{3.3}$$

where z is the coordinate along the thickness direction and $z = 0$ is the mid-plane of the plate. w is the deflection, and ψ_x and ψ_y are rotations. Considering an isotropic plate with thickness h and density ρ, the equations of motion in vector form can be expressed as [198]

$$D_s\left\{\nabla\nabla\cdot\mathbf{\Psi} - \frac{1-\upsilon}{2}\nabla\times\nabla\times\Psi\right\}\rho + \kappa^2\mu h(\nabla w - \mathbf{\Psi}) - \rho I\ddot{\mathbf{\Psi}} = \mathbf{m}, \tag{3.4}$$

$$\kappa^2\mu h\nabla\cdot(\nabla w - \mathbf{\Psi}) - \rho h\ddot{w} = -q, \tag{3.5}$$

where $\kappa = \pi/\sqrt{12}$ is the shear correction factor, which can be used to take into account the low frequency behavior [79]. D_s donates the plate bending stiffness and $D_s = EI/(1-\upsilon)^2$. E is the Young's modulus and μ is the shear modulus υ is the Poisson's ratio and I is the moment of inertia, and $I = h^3/12$. $\mathbf{m} = [m_x, m_y]^T$ and $\mathbf{\Psi} = [\psi_x, \psi_y]^T$ denote a vector consists of the rotations and area density of the external bending moments, respectively. $\ddot{\mathbf{\Psi}} = \partial^2\mathbf{\Psi}/\partial t^2$ where t is time. It has been shown that the general solution of harmonic motion without loads is of the form [246]

$$w = \mathrm{Re}\left[W(x,y)e^{-i\omega t}\right], \tag{3.6}$$
$$(\psi_x, \psi_y) = \left(\mathrm{Re}\left[\psi_x(x,y)e^{-i\omega t}\right], \mathrm{Re}\left[\psi_y(x,y)e^{-i\omega t}\right]\right), \tag{3.7}$$

where $i = \sqrt{-1}$ is the imaginary unit and ω is the angular frequency, and

$$W(x,y,t) = W_1(x,y,t) + W_2(x,y,t), \tag{3.8}$$
$$\mathbf{\Psi}(x,y,t) = \lambda_1\nabla W_1(x,y,t) + \lambda_2\nabla W_2(x,y,t) + \nabla\times(\mathbf{e}_z V(x,y,t)), \tag{3.9}$$

where λ_1 and λ_2 are the frequency dependent coefficients. $\mathbf{e}_z$ is the unit vector in the plate thickness direction. W_1, W_2, and V each satisfy the Helmholtz equations [246]

$$(\nabla^2 + k_1^2)W_1 = 0, \tag{3.10}$$
$$(\nabla^2 + k_2^2)W_2 = 0, \tag{3.11}$$
$$(\nabla^2 + k_3^2)V = 0, \tag{3.12}$$

where k_1 is a real wavenumber representing the flexural wave, k_2 and k_3 are purely imaginary wave numbers representing evanescent waves when the frequency is below the cut-off frequency $\omega_c = \kappa\sqrt{\mu h/\rho I}$.

The wavenumbers k_1, k_2, and k_3 can be determined by [198]

$$k_{1,2}^2 = \omega^2\left(B \pm \sqrt{B^2 - C(\omega)}\right), \tag{3.13}$$

$$k_3^2 = \sqrt{2\frac{\rho I\omega^2 - \kappa^2\mu h}{D_s(1-\nu)}}, \tag{3.14}$$

where

$$B = \frac{1}{2}\left(\frac{\rho I}{D_s} + \frac{\rho}{\kappa^2 \mu}\right), \quad C(\omega) = \frac{\rho I}{D_s}\frac{\rho}{\kappa^2 \mu} - \frac{\rho h}{D_s}\frac{1}{\omega^2}. \tag{3.15}$$

The parameters λ_1 and λ_2 can be calculated by

$$\lambda_j = -1 + \frac{k_b^2}{k_j^2}, \quad j = 1, 2. \tag{3.16}$$

The ratio of amplitudes of Equations (3.9) and (3.8) is

$$\gamma_i(k,\omega) = \frac{\psi_i(k,\omega)}{W(k,\omega)} = 1 - \frac{\rho}{\kappa^2 \mu}\frac{\omega^2}{k_i^2}. \tag{3.17}$$

The phase velocity $c_p = \omega/k$ and group velocity $c_g = d\omega/dk$ are readily obtained from Equations (3.13)–(3.15) as follows

$$c_p(\omega) = \frac{1}{\sqrt{B \pm \sqrt{B^2 - C(\omega)}}}, \tag{3.18}$$

$$c_g(\omega) = c_p(\omega)^2 \left[c_p(\omega) - \omega \frac{dc_p(\omega)}{d\omega}\right]. \tag{3.19}$$

In the cylindrical coordinate system, the deflection of an incident wave at frequency ω can be written as [253]

$$W^{(i)}(r,\theta) = \sum_{n=0}^{\infty} \epsilon_n\, i^n J_n(k_1 r)\cos(n\theta), \tag{3.20}$$

where the superscript i denotes parameters pertaining to the incident wave. $J_n(\cdot)$ is the n-th order Bessel function of the first kind, and

$$\epsilon_n = \begin{cases} 1, & n = 0, \\ 2, & n \geq 1. \end{cases} \tag{3.21}$$

3.3 LAMB WAVE DIFFRACTION TOMOGRAPHY

The Lamb wave diffraction tomography presented in this chapter relies on using Mindlin's plate theory to represent wave scattering by damage and take into account the diffraction effect. In this section, the damage is modeled as a finite region Σ of inhomogeneity with different isotropic plate properties such as the plate bending stiffness D_s, the shear stiffness $\kappa^2 \mu h$, the rotary inertia ρI, and the transverse inertia ρh varying with position $\mathbf{x}$. Using the plate properties outside the inhomogeneity, the corresponding plate properties of the inhomogeneity region can be expressed as [196,198]

$$D_s^{\Sigma} = [1 + s_1(\mathbf{x})] D_s, \tag{3.22}$$

$$(\kappa^2 Gh)^{\Sigma} = [1 + s_2(\mathbf{x})] \kappa^2 \mu h, \tag{3.23}$$

$$(\rho I)^{\Sigma} = [1 + s_3(\mathbf{x})] \rho I, \tag{3.24}$$

$$(\rho h)^{\Sigma} = [1 + s_4(\mathbf{x})] \rho h, \tag{3.25}$$

where $\mathbf{x} = (x_1, x_2)$. s_1, s_2, s_3, and s_4 are variations for characterizing the inhomogeneity, which vanish outside the finite region Σ and nonzero within it. These variations can be used to represent different types of damage. For example, a variation of D_s can be due to a variation of Young's modulus due to cracking or a variation of h due to corrosion. However, the density ρ is usually assumed to have the same value inside and outside the finite region Σ in the isotropic plate. In this chapter, the damage is considered as plate thickness variation, which is also assumed as symmetrical with respect to the mid-plane of the plate and, hence, to avoid the mode coupling effect. Therefore, the variations have the relation $3s_1 = s_2 = 3s_3 = s_4$.

3.3.1 INTEGRAL EQUATION FOR SCATTERING

The total displacement $\mathbf{u}$ obtained from measurement is a superposition of incident wave displacement $\mathbf{u}^I$ and scattered wave displacement $\mathbf{u}^S$, and can be expressed as

$$\mathbf{u} = \mathbf{u}^I + \mathbf{u}^S. \tag{3.26}$$

The scattered field, with the consideration of the Fourier transform over time of all plate theory variables, can be represented by

$$\hat{u}_i^S(\mathbf{x}, \omega) = \iint_{\Sigma} \hat{f}_j(\xi) \hat{g}_{ij}(\mathbf{x} - \boldsymbol{\xi}) d\xi d\eta, \tag{3.27}$$

where the symbol ^ means the response is the plate-theory body force. ξ is an arbitrary point within the finite region Σ. $\hat{g}_{ij}$ is Green's function and has been derived in [198]. Using integration by parts and Gauss's theorem, a more convenient representation can be obtained as [199]

$$\hat{u}_i^S(\mathbf{x}, \omega) = \iint_{\Sigma} \left\{ s_1 D_s \hat{\Gamma}_{\beta\alpha} \hat{g}_{i\alpha,\beta} + \left[s_2 \kappa^2 \mu h (\hat{w}_{,\alpha} - \hat{\Omega}_{\alpha}) + s_3 \omega^2 \rho I \hat{\Omega}_{\alpha} \right] \hat{g}_{i\alpha} \right.$$
$$\left. + s_2 \kappa^2 \mu h (\hat{w}_{,\alpha} - \hat{\Omega}_{\alpha}) \hat{g}_{i3,\alpha} + s_4 \omega^2 \rho h \hat{g}_{i3} \right\} d\xi d\eta, \tag{3.28}$$

where $s_n (n = 1, ..., 4)$ are non-zero values for $\mathbf{x} \in \Sigma$ but vanish for $\mathbf{x} \notin \Sigma$. $\hat{\Gamma}_{\beta\alpha}$ are the plate-theory strains. $\hat{w}_{,\alpha}$ and $\hat{\Omega}_{\alpha}$ are the plate-normal displacement and rotary deflections.

3.3.2 BORN APPROXIMATION

Assuming the damage is a weak inhomogeneity and using Born approximation, $\hat{\mathbf{u}} = \hat{\mathbf{u}}^I + \hat{\mathbf{u}}^S$ can be approximated by $\hat{\mathbf{u}} \approx \hat{\mathbf{u}}^I$, and the following approximation can be achieved

$$\hat{u}^S(\mathbf{x}, \omega) \approx \hat{u}^B(\mathbf{x}, \omega). \tag{3.29}$$

The scattered field can be approximated as

$$\hat{u}_i^B(\mathbf{x}, \omega) = \iint_\Sigma \Big\{ s_1 D_s \hat{\Gamma}^I_{\beta\alpha} \hat{g}_{i\alpha,\beta} + \left[s_2 \kappa^2 \mu h (\hat{w}^I_{,\alpha} - \hat{\Omega}^I_\alpha) + s_3 \omega^2 \rho I \hat{\Omega}^I_\alpha \right] \hat{g}_{i\alpha} \\ + s_2 \kappa^2 \mu h (\hat{w}^I_{,\alpha} - \hat{\Omega}^I_\alpha) \hat{g}_{i3,\alpha} + s_4 \omega^2 \rho h \hat{g}_{i3} \Big\} d\xi \, d\eta. \tag{3.30}$$

By restricting the frequency below the cut-off frequency, i.e., only the first mode is the propagating mode, using the far-field approximation for Hankel function in the Green's functions and assuming the incident wave is a plate wave, the far-field scattered wave can be simplified as [199]

$$\hat{w}^S(r, \theta; \omega) = -\frac{i}{4} \sqrt{\frac{2}{\pi k_1 r}} e^{i(k_1 r - \pi/4)} A(\theta, \alpha; \mathbf{0}) + O(1/r), \tag{3.31}$$

and hence we can define the plate wave scattering amplitude as

$$A(\theta, \alpha; \mathbf{0}) = \lim_{k_1 r \to \infty} \frac{\hat{w}^S(r, \theta; \omega)}{\frac{i}{4} \sqrt{\frac{2}{\pi k_1 r}} e^{i k_1 r} e^{-i\pi/4}}, \tag{3.32}$$

where $\hat{w}^S(r, \theta; \omega)$ is the scattered wave field along direction θ due to an incident plane wave of unit strength in the direction α.

3.3.3 IMAGING INHOMOGENEITIES

Lamb wave diffraction tomography provides a relation between the 2D spatial Fourier transform of the inhomogeneity variation and the plate wave scattering amplitude. The inverse Fourier transform in the wavenumber space can reconstruct the inhomogeneity variation. This inversion in the **k**-space is integral over the source and receiver angles. The reconstructed value of the plate thickness reduction $s(\mathbf{x})$ is

$$s(\mathbf{x}) = \frac{1}{8\pi^2} \int_0^{2\pi} \int_0^{2\pi} \frac{A(\theta, \alpha; \mathbf{0})}{q_h(\theta - \alpha)} \times |\sin(\theta - \alpha)| e^{i k_1 (\theta - \alpha) \cdot \mathbf{x}} d\theta d\alpha, \tag{3.33}$$

where $|\sin(\theta - \alpha)|$ is the Jacobian for the change in integration variables from $(k_x^S - k_x^I, k_y^S - k_y^I)$ to (θ, α). $q_h(\theta - \alpha)$ is a shape function used to characterize the scattering pattern for a point-scatterer. For plate thickness reduction, it is $q_h = 3q_1 + q_2 + 3q_3 + q_4$ with the following expressions

$$q_1(\Delta\theta) = \frac{\gamma_1 k_1^2}{(k_1^2 - k_2^2)}[\cos^2\Delta\theta + \nu \sin^2\Delta\theta], \tag{3.34}$$

$$q_2(\Delta\theta) = \frac{\kappa^2 \mu h(1-\gamma_1)^2}{\gamma_1 D_s(k_1^2 - k_2^2)} \cos\Delta\theta, \tag{3.35}$$

$$q_3(\Delta\theta) = -\frac{\gamma_1 \rho \omega^2}{D_s(k_1^2 - k_2^2)} \cos\Delta\theta, \tag{3.36}$$

$$q_4(\Delta\theta) = -\frac{h\rho\omega^2}{\gamma_1 D_s k_1^2 (k_1^2 - k_2^2)}, \tag{3.37}$$

where γ is the wave amplitude ratio between the first and second branches of the wave mode. $\Delta\theta = \theta - \alpha$ or $\Delta\theta = \varphi_i - \varphi_j + \pi$.

Equation (3.33) can be generalized by two steps. In the first step, the incident wave at **x** due to a vertical point force $\hat{F}(\omega)$ generated by an actuator at $\mathbf{X}_S = (x_s, y_s)$ can be expressed as

$$\hat{w}^I(\mathbf{x}, \omega; \mathbf{X}_S) = \hat{F}(\omega) G(\mathbf{x}, \omega; \mathbf{X}_S), \tag{3.38}$$

where $G(\mathbf{x}, \omega; \mathbf{X}_S)$ is the plate wave dynamic Green's function, which is the response at **x** due to a unit excitation force at $\mathbf{X}_S$, but it is expressed by notation in terms of the position **x**, the frequency ω, and the excitation location $\mathbf{X}_S$. Figure 3.1 shows the schematic diagram for the Lamb wave diffraction tomography. R is the radius of the circular transducer network.

In this chapter, the analytical plate wave dynamic Green's function given by [166, 198], is used, and it is defined as

$$G(\mathbf{X}_r, \omega; \mathbf{x}) = \frac{1}{D_s(k_1^2 - k_2^2)\gamma_1} \frac{i}{4} \sqrt{\frac{2}{\pi k_1 r}} e^{ik_1 \rho_r} e^{-i\frac{\pi}{4}}. \tag{3.39}$$

The scattered field measured by a sensor at location $\mathbf{X}_r = (x_r, y_r)$ can be then expressed as

$$\hat{w}^S(x_r, y_r; \omega) = A(\varphi_r, \varphi_s - \pi; \mathbf{x}) \hat{w}^I(\mathbf{x}, \omega; \mathbf{X}_s) \left(\frac{i}{4} \sqrt{\frac{2}{\pi k_1 \rho_r}} e^{ik_1 \rho_r} e^{-i\pi/4} \right). \tag{3.40}$$

In the second step, the following translation equation relates the plane wave scattering amplitude relative to an origin at **x** to the plate wave scattering amplitude relative to the origin at **0**

$$A(\theta, \alpha; \mathbf{x}) = A(\theta, \alpha; \mathbf{0}) e^{jk_1(\theta - \alpha)\cdot \boldsymbol{x}}. \tag{3.41}$$

With these two steps, Equation (3.33) can be rewritten as

$$s(\mathbf{x}) = \frac{1}{8\pi^2 D_s \gamma_1 (k_1^2 - k_2^2)} \int_0^{2\pi} \int_0^{2\pi} \frac{K_{rs} |\sin(\varphi_r - \varphi_s)|}{G(\mathbf{X}_r, \mathbf{x}) G(\mathbf{x}, \mathbf{X}_s) q_h(\varphi_r - \varphi_s + \pi)} d\varphi_r d\varphi_s, \tag{3.42}$$

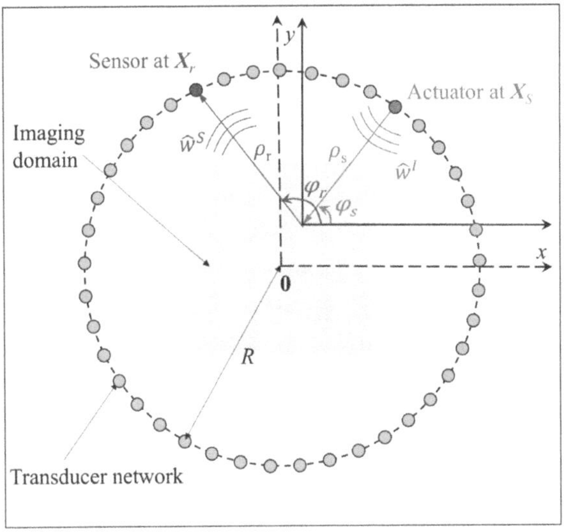

Figure 3.1 Schematic diagram of the transducer network and imaging domain in Lamb wave diffraction tomography [166].

where K_{rs} is the scattered field data normalized by the amplitude of the actuation force $\hat{F}(\omega)$ at the center frequency ω of the excitation signal

$$K_{rs} = \frac{\hat{w}^S(\mathbf{X}_r, \omega; \mathbf{X}_s)}{\hat{F}(\omega)}. \tag{3.43}$$

The integrals in Equation (3.42) can be numerically evaluated using trapezoidal quadrature rule as

$$s(\mathbf{x}) = \mathrm{Re}\left[\frac{1}{8\pi^2 D_s \gamma_1 (k_1^2 - k_2^2)} \sum_{j=1}^{N} \sum_{i=1}^{N} \frac{K_{ij} |\sin(\varphi_i - \varphi_j)|}{G(\mathbf{X}_i, \mathbf{x}) G(\mathbf{x}, \mathbf{X}_j) q_h(\varphi_i - \varphi_j + \pi)} \times \Delta\varphi_i \Delta\varphi_j\right], \tag{3.44}$$

where $\Delta\varphi_i = (\varphi_{i+1} - \varphi_{i-1})/2$ and $\varphi_0 = \varphi_N - 2\pi$. K_{ij} is defined as the multi-static data matrix defined in Equation (3.43). i and j represent the sensor and actuator locations, respectively. N is the total number of transducers. φ is the polar angle relative to the imaging point $\mathbf{x}$ as origin. Therefore, the thickness reduction of the damage can be obtained using Equation (3.44).

3.4 NUMERICAL CASE STUDIES

In this section, a series of numerical case studies are presented to evaluate the performance of the Lamb wave diffraction tomography. The 3D FE simulation is used

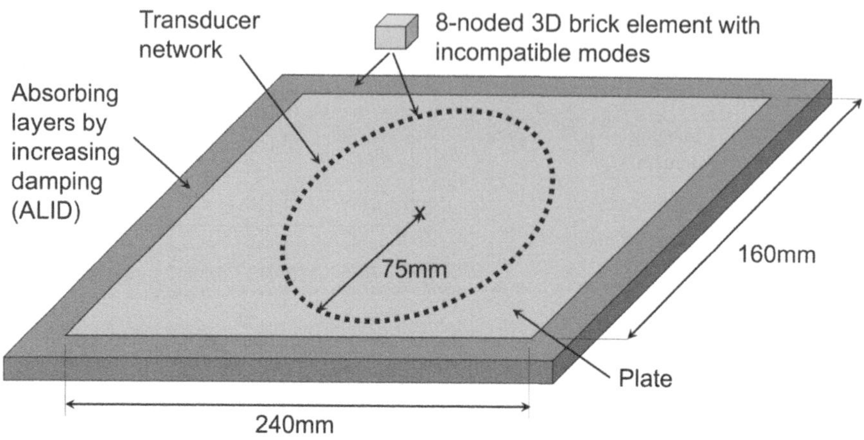

Figure 3.2 Schematic diagram of the 3D finite element model [166].

to simulate the Lamb wave propagation and scattering at a damage in an aluminum plate. The FE simulation is carried out using a commercial FE software, ABAQUS. The dimensions of the plate are 240mm×240mm×1.6mm. It is assumed that the plate has Young's modulus $E = 69$GPa, density $\rho = 2760$kg/m^3, and Poisson's ratio $v = 0.33$. Eight-noded 3D reduced integration solid brick element (C3D8I) with incompatible modes is used to model the plate, in which each node has three degrees of freedom (DoFs). The fundamental anti-symmetric mode (A_0) of the Lamb wave is used as the incident wave. The excitation signal is a 200 kHz narrow-band five-cycle sinusoidal tone burst modulated by a Hann window. A circular transducer network consisting of 128 transducers is used for a sequential scan, by which one of the transducers is used for actuating the A_0 Lamb wave while the rest of the transducers are used for sensing. As shown in Figure 3.2, a 160mm×240 mm plate is modeled. The radius of the transducer network is ten wavelengths of the excited A_0 Lamb wave, i.e., $10\lambda = 75$ mm. Absorbing layers by increasing damping (ALID) [57] are adopted at four edges of the plate to minimize the computational cost of the FE simulations so that it can present a large plate. The damping values in the absorbing layer are determined by gradually varying the mass-proportional Rayleigh damping until the wave reflected from boundaries is eliminated.

The A_0 Lamb wave is generated by applying out-of-plane nodal displacement to the FE nodes at the surface of the plate, which is assumed to be covered by the transducer. The Lamb wave is captured by the out-of-plane nodal displacement of the nodes located at the surface of the plate. To ensure the accuracy of simulating the wave propagating and scattering at the damage, the solid brick elements have an in-plane square shape with dimensions 0.4 mm×0.4 mm, which ensures there are at least 12 nodes per wavelength. The plate is modeled by six layers of elements in the thickness direction, in which the thickness of the top and bottom layers is 0.08 mm

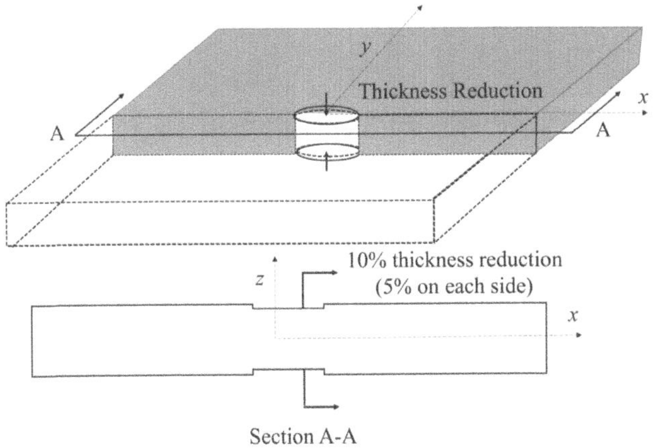

Figure 3.3 Cross-section view of the circular thickness reduction damage [166].

while the thickness of the layers in between is 0.36 mm.

The damage is modeled by reducing the thickness of the plate, and it is assumed that the thickness reduction is symmetric about the mid-plane of the plate, and hence there are no other wave modes generated at the damage due to the mode-conversion effect. In this study, a 10% thickness reduction is used for all the damages in this chapter, and three different shapes of damage are considered: Case 1: a circular-shaped damage, Case 2: a cross-shaped damage, and Case 3: a square-shaped damage. Figure 3.3 shows a schematic diagram of the circular-shaped damage. The results of these case studies were published in [166], and part of these results are presented below.

The dynamic simulation is solved by ABAQUS/Explicit and the increment time step is automatically controlled by ABAQUS based on the element size and material properties of the plate. Each of the transducers is used as both the actuator and sensor to excite and measure the wave signal. A sequential scan is carried out by exciting one of the transducers to generate the A_0 Lamb wave while the rest of the transducers are used for measuring the wave signals.

To extract the scattered wave field, two simulations are carried out: one for the intact model and the other for the model with the damage. Using the baseline subtraction method, the scattered wave field is obtained by subtracting the signal of the intact model from the signals of the model with the damage. The scattered wave field obtained from each of the transducers at 200 kHz is used to construct a single row of the multi-static data matrix K_{ij} in Equation (3.44), and the process is repeated until a $N \times N$ matrix is obtained, where N is the number of transducers.

Case 1 considers a 12 mm diameter circular thickness reduction located at the center of the plate. The diameter of this circular damage is roughly equal to 1.5 wavelengths of the excited A_0 Lamb wave. As it is a circular-shaped damage and

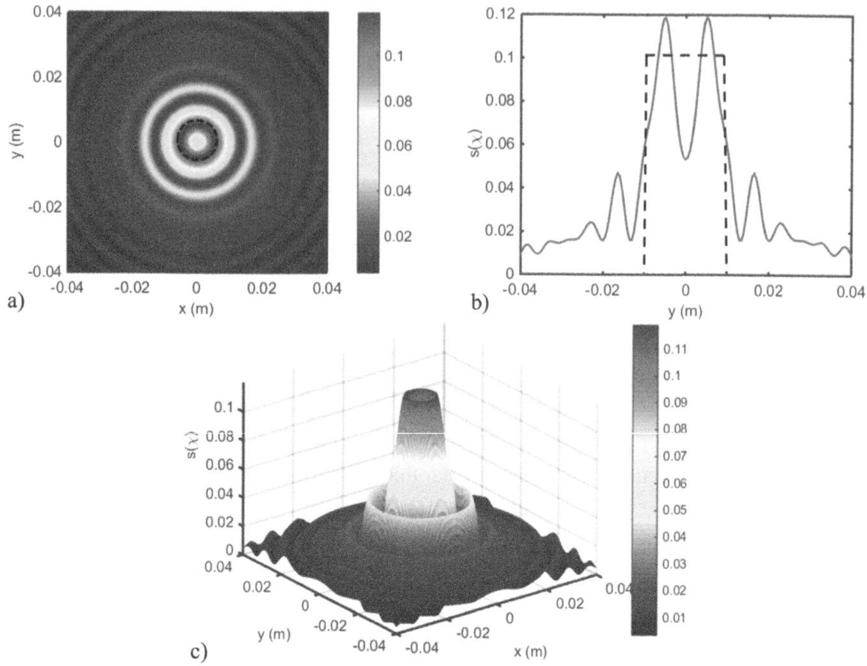

Figure 3.4 Reconstructed image of the circular damage: (a) 2D variation of the damage along the surface of the plate, (b) damage profile along the x-axis, and (c) 3D variation of damage (dashed lines: true damage geometry and thickness reduction) [166].

located at the center of the plate, which is the center of the transducer network, the simulation results of exciting one of the transducers can be replicated to represent the results of exciting other transducers in the sequential scan, and hence, it only requires one simulation. It should be noted that while the computational cost of the FE simulation can be significantly reduced, care must be taken in assembling the multi-static data matrix, as the order of the elements in each row of the matrix depends on the order and location of the transducer excited. Using the Lamb wave diffraction tomography, the reconstructed damaged image is shown in Figure 3.4. Figure 3.4a shows the 2D variation of the damage along the surface of the plate, and Figure 3.4b is the damage profile. The dashed lines are the true damage geometry and thickness reduction. Figure 3.4 shows the 3D variation of the damage. The results show that the prediction of the thickness reduction is accurate, but it only has a reasonable reconstruction of the damage shape.

Case 2 considers the cross-shaped thickness reduction located at the center of the plate. The size of the cross-shaped is 12 mm × 12 mm. The width of the intersecting bars is 2 mm. Since the cross-shape is symmetric in each quadrant as shown by the dashed line in Figure 3.5, the FE simulation cannot be simplified as that in Case 1.

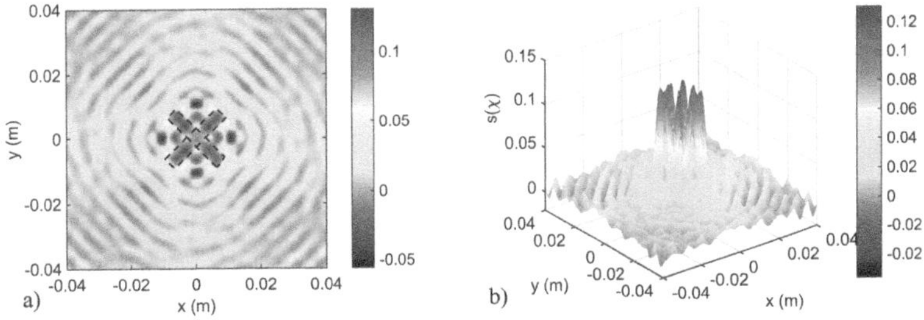

Figure 3.5 Reconstructed image of the crossed-shaped damage: (a) 2D variation of the damage along the surface of the plate, and (b) 3D variation of the damage (dashed lines: true damage geometry and thickness reduction) [166].

However, a certain level of simplification can still be achieved by which we simulate the excitation of exciting 32 transducers sequentially. The simulation results of the other 96 transducers are obtained by replicating the simulation results with careful rearrangement of the obtained simulation results. The reconstructed damage image is shown in Figure 3.5. The predicted thickness reduction is accurate but there are some discrepancies between the reconstructed shape and the actual geometry of the damage. As shown in Figure 3.5, there are additional peaks between the top, bottom, left, and right sides of the cross-shaped damage. This is due to the irregular meshing at the corners of the cross-shaped damage.

In Case 3, the square-shaped damage is considered. The size of the square-shaped damage is 12 mm × 12 mm and is located at the center of the plate. Similarly to cross-shaped damage, square-shaped damage is also symmetric in each quadrant; thus, the same simulation simplification is adopted. The predicted thickness reduction is shown in Figure 3.6. As shown in the 2D and 3D variations of the damage, the predicted thickness reduction and geometry are accurate.

3.5 SUMMARY

In this chapter, a Lamb wave diffraction tomography developed on the basis of Mindlin's plate theory is presented. The key steps in the formulation of Mindlin's plate theory are first presented, and then the modeling of the damage, the formulation of the integral equation for scattering, the Born approximation, and the multi-static data matrix are presented. Although an analytical Green's function is employed in this chapter, the framework can employ the numerically computed Green's function to account for the effects of structural complexities. The 3D FE simulation is used to generate the simulation data to demonstrate the reconstruction of the damaged image. Three cases are presented in this chapter, which consider circular, cross-shaped, and square-shaped thickness reduction. The thickness reduction considered in these cases is only 10% of the plate thickness. The reconstructed damage image can

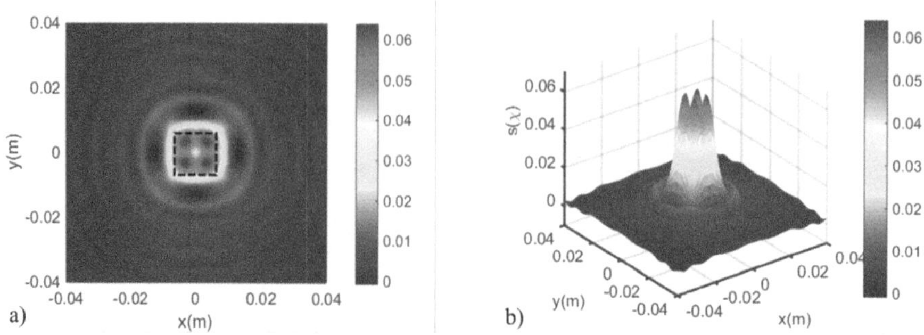

Figure 3.6 Reconstructed image of the square-shaped damage: (a) 2D variation of the damage along the surface of the plate, and (b) 3D variation of the damage (dashed lines: true damage geometry and thickness reduction) [166].

provide satisfactory performance in predicting the thickness reduction and geometry of the damage.

3.6 EXERCISES

1. What assumptions are made in the Lamb wave diffraction tomography presented in this chapter?
2. What are the limitations of using the analytical plate wave dynamic Green's function in Lamb wave diffraction tomography?
3. What are the advantages of using the numerically computed Green's function in Lamb wave diffraction tomography?
4. The relationship of the variation presented in Section 3.3 is used for plate thickness reduction. Can you describe a relationship of variation for other types of damage?

4 Tomography Based on Full Waveform Inversion

4.1 INTRODUCTION

It has been demonstrated that the goal of defect detection using the Mindlin plate theoretical framework can be achieved with certain simplifications and idealizations. The most significant of these simplifications is neglecting mode coupling caused by realistic forms of defects. The interaction between incident waves and defects can be adequately captured within plate theory by modeling defects as inhomogeneities. Therefore, laminar defects can be modeled as flexural inhomogeneities within the Mindlin plate theory. However, the use of the Born approximation in this method limits its imaging capabilities to weakly scattering objects.

In the remaining chapters of this book, a quantitative ultrasonic imaging technique based on full waveform inversion (FWI) is presented. This technique plays a crucial role in corrosion mapping and quality control. For corrosion mapping, the mono-parameter FWI in ultrasonic guided wave tomography (UGWT) is proposed to reconstruct the remaining thickness of corrosion damages in plate-like and pipe structures. This reconstruction is achieved by employing velocity-to-thickness mapping using dispersion curves, as covered in Chapters 3 to 7. Building upon this foundation, mono-parameter FWI in UGWT is further extended to reconstruct Young's modulus and account for anisotropy in laminated composites, as presented in Chapter 8. In Chapter 9, improved FWI in ultrasonic bulk wave tomography (UBWT) employs various descent methods to address the optimization problem. Finally, in Chapter 10, the quality control application involves developing multi-parameter FWI in UBWT for the simultaneous reconstruction of velocity and density in metal parts.

In this chapter, the mono-parameter FWI in UGWT is introduced to reconstruct the velocity map from full-view configurations and then convert this to thickness based on dispersion curves. The theory of the mono-parameter FWI, based on the finite difference (FD) frequency domain method, is introduced in Section 4.2, which includes forward modeling and inversion. Section 4.3 presents numerical simulations using a simplified acoustic model implemented with the FD method, as well as a more realistic elastic model implemented with the finite element (FE) method. The choice of frequencies and the calibration method are also discussed in this section. The experimental setup and data processing are demonstrated in Section 4.4. The performance of the mono-parameter FWI, including its sensitivity, is evaluated in Section 4.5 through the analysis of two representative examples. Finally, the discussion of computational aspects related to the mono-parameter FWI is followed in Section 4.6.

DOI: 10.1201/9781003667926-4

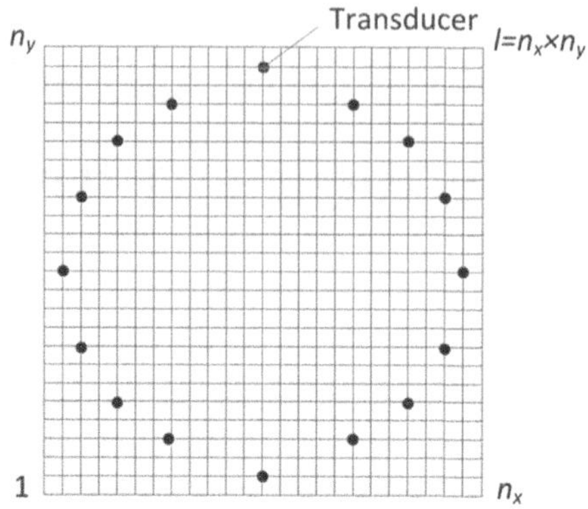

Figure 4.1 A schematic discrete representation of the $l = n_x \times n_y$ finite difference grid points for forward modeling [169].

4.2 THEORY

4.2.1 FORWARD MODELING IN THE SPACE-FREQUENCY DOMAIN

The frequency domain 2D acoustic wave equation in a constant density medium can be expressed as follows:

$$(\nabla^2 + K^2)p(x, y, \omega) = -s(x, y, \omega), \tag{4.1}$$

where $p(x, y, \omega)$ is the pressure (or displacement) wavefield, $s(x, y, \omega)$ being the source, ω being the angular frequency, and $K(x, y, \omega) = \omega / v(\omega)$ is the wavenumber linked with the phase velocity v.

Equation (4.1) can be discretized using the FD method with the mixed-grid approach [87, 102, 229]. Repeating this method at all grid points leads to a large system of linear equations. To solve these equations, they are rewritten in matrix form

$$\mathrm{AP} = \mathrm{S} \quad or \quad \mathrm{P} = \mathrm{A}^{-1}\mathrm{S}, \tag{4.2}$$

where the complex-valued impedance matrix, also known as the forward modeling operator, is represented as $\mathrm{A} = \nabla^2 + K^2$. This matrix depends on the frequency and physical properties of the media. To illustrate the discretization, Figure 4.1 provides a visual representation. The pressure wavefield is calculated at $l = n_x \times n_y$ grid points on the 2D regular grid, where n_x and n_y denote the number of grids in the horizontal and vertical directions, respectively. The pressure wavefield P and the source term S at one frequency are mapped into $l \times 1$ column vectors. The complex-valued impedance matrix A is an $l \times l$ matrix.

Equation (4.2) is typically solved using the direct method of LU factorization. By decomposing matrix A through LU factorization, the factored A matrix can be reused to solve the forward problem for other source vectors. This reusability of the factorized matrix is particularly advantageous when handling multiple sources, as it enables efficient computation of problems involving multiple sources. This efficiency is vital in the iterative solutions of inverse problems, where forward solutions for both real and "virtual" sources (to be explained subsequently) are required in each iteration. Consequently, Equation (4.2) can be represented as follows:

$$\mathrm{LU}[\mathrm{P_1P_2}\cdots\mathrm{P_n}] = [\mathrm{S_1S_2}\cdots\mathrm{S_n}], \tag{4.3}$$

where n is the number of sources.

The use of absorbing boundaries along the edges of the model can effectively prevent reflections at those boundaries [19, 45, 87]. This technique allows for a reduction in the model size by effectively truncating the computational domain, while still preserving the accuracy of the simulation.

4.2.2 INVERSE PROBLEM IN THE SPACE-FREQUENCY DOMAIN

Figure 4.2 outlines the structure of the FWI algorithm. The multi-resolution nature of the reconstruction is controlled through loop 1, which traverses several discrete frequencies, starting from low and progressing to higher frequencies. At each frequency, several iterations are conducted, as indicated by loop 2. After each iteration at a given frequency, a new model parameter is calculated and set as the initial model for the subsequent iteration, until a predefined maximum number of iterations, chosen to be between 20 and 40 for this study, is reached. At each iteration, the residual data (the difference between computed results from the current model and the observed data from experiments) are minimized in the sense of least squares. The velocity map obtained from the final iteration at the current frequency becomes the initial model for the next frequency. This process repeats until the convergence criterion, which will be detailed later, is met at the highest frequency. The goal of the inversion is to create a set of model parameters (m) that can reproduce the observations through forward modeling. These model parameters consist of the values of the squared slowness [220]. The theory of the FWI has been introduced in detail in seismology [164, 189, 212]; hence, only the principal equations are presented here.

The weighted least squares norm of the objective function, i.e., the L_2 norm of the data residuals, is defined as

$$C^{(k)}(\mathrm{m}) = \frac{1}{2}\Delta\mathrm{d}^{\dagger}\mathrm{W}_d\Delta\mathrm{d}, \tag{4.4}$$

where $\Delta\mathrm{d} = d_{calc}^{(k)} - d_{obs}$ is the data residual (difference between data $d_{calc}^{(k)}$ calculated in the current model $\mathrm{m}^{(k)}$ and the observed data d_{obs}). The superscript † represents the transposed conjugate. W_d is a weighting vector that is used in the data residual to scale the relative contribution of each component in the inversion [151], and k is the iteration number.

The inverse problem is to minimize the objective function, and it can be computed by using the gradient

$$G^{(k)} = Re\{\mathrm{J}^t\mathrm{W}_d\Delta\mathrm{d}^*\}, \tag{4.5}$$

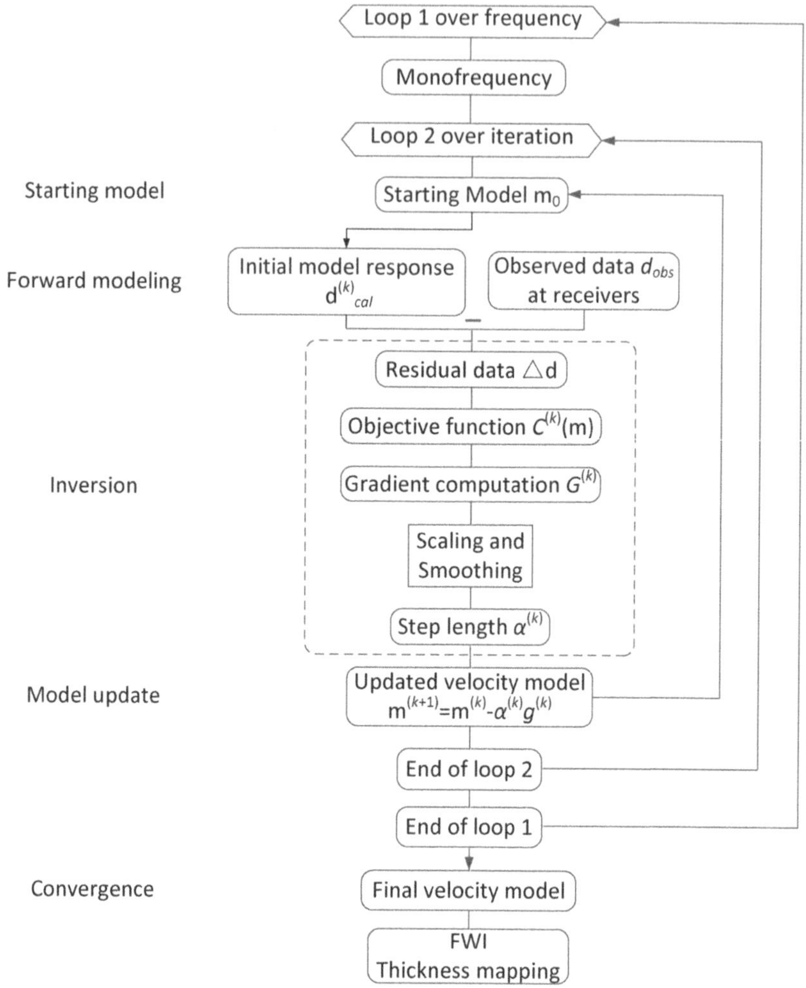

Figure 4.2 Structure of the mono-parameter FWI algorithm [170].

where J^t is the transpose of the Fréchet derivative matrix (i.e., Jacobian matrix, $J = \frac{\partial P}{\partial m}$). Δd^* is the conjugate of the data residual, and *Re* is the real part of a complex number [189].

Furthermore, to derive the expression for any of partial derivatives $\frac{\partial P}{\partial m}$ in Equation (4.5), the partial derivatives of both sides of the forward equation (4.2) are taken with respect to the *i*th model parameter m_i

$$A\frac{\partial P}{\partial m_i} = -\frac{\partial A}{\partial m_i}P. \tag{4.6}$$

A "virtual" source term $f^{(i)} = -\frac{\partial A}{\partial m_i} P$ can be introduced as

$$\frac{\partial P}{\partial m_i} = A^{-1} f^{(i)}. \tag{4.7}$$

An analogy between Equations (4.2) and (4.7) indicates that partial derivatives $\frac{\partial P}{\partial m}$ are the solutions of a new forward model of the "virtual" source $f^{(i)}$. Thus, the link between the gradient vector (including the partial derivatives, $\frac{\partial P}{\partial m}$) and new forward modeling can be established.

Since the equation similar to Equation (4.7) can be generated for any value of i, all partial derivatives can be given by the following matrix equation:

$$J = \left[\frac{\partial P}{\partial m_1} \frac{\partial P}{\partial m_2} \cdots \frac{\partial P}{\partial m_q} \right] = A^{-1} [f^{(1)} f^{(2)} \cdots f^{(q)}], \tag{4.8}$$

where q is the number of model parameters, $q \leq l$. The computation of elements of J requires solving q forward-propagation problems using virtual sources.

Thus, to obtain the gradient using Equation (4.5), it is unnecessary to calculate elements of partial derivatives J directly. By substituting Equation (4.8) and $f^{(i)}$ into Equation (4.5), we obtain:

$$G^{(k)} = Re \left\{ -P^t \left[\frac{\partial A^t}{\partial m_i} \right] A^{-1t} W_d \Delta d^* \right\}. \tag{4.9}$$

Since A^{-1} is symmetric in the acoustic problem, i.e., $[A^{-1}]^t = A^{-1}$, Equation (4.9) can be given by

$$G^{(k)} = Re \left\{ -P^t \left[\frac{\partial A^t}{\partial m_i} \right] A^{-1} W_d \Delta d^* \right\}, \tag{4.10}$$

where $A^{-1} W_d \Delta d^*$ is defined as backward propagated wavefield. The matrix $\frac{\partial A^t}{\partial m_i}$ can be easily computed using the coefficients of the matrix A. Based on the reciprocity principle, the gradient is computed by zero-lag convolution of the forward-modeled wavefield P with the backward-propagated residual wavefield $A^{-1} W_d \Delta d^*$ (i.e., the adjoint-state method). This means that only two forward models per source are required. The first forward problem is to obtain the wavefield P for a source location. The second forward problem calculates the backward-propagated residual wavefield using a "composite" source formed by assembling data residuals.

To obtain reliable and stable results, the regularization and scaling are applied to the gradient Equation (4.10). It can be modified to

$$G^{(k)} = (diag H_a + \varepsilon I)^{-1} \mathscr{G}_{2D} Re \left\{ -P^t \left[\frac{\partial A^t}{\partial m_i} \right] A^{-1} W_d \Delta d^* \right\}, \tag{4.11}$$

where $diag H_a = diag Re\{J^t W_d \Delta J^*\}$ indicates the diagonal elements of weighted approximate Hessian H_a; ε is the damping factor and $\mathscr{G}_{2D}$ is the spatial smoothing operator.

The diagonal of H_a provides a proper preconditioner for the gradient which scales the tomographic model. The diagonal element of H_a is the scalar product of two partial derivative wavefields. Furthermore, as the scatterers are removed from the source position, the amplitudes of the partial derivative wavefields and the corresponding elements of H_a will decrease. Dividing data residuals by these squared amplitude terms is equivalent to removing effects of geometrical amplitudes decreasing partial derivative wave-fields from data residuals [151]. The damping factor ε is added to the diagonal elements of H_a, which can stabilize the inversion process because the H_a may be ill-conditioned or singular. The smoothing operator $\mathscr{G}_{2D}$ is in the form of a 2D Gaussian spatial filter in this book, and its correlation length is adapted to inverted frequency components.

Finally, the model parameter vector is updated iteratively according to

$$\mathrm{m}^{(k+1)} = \mathrm{m}^{(k)} - \alpha^{(k)} g^{(k)} \tag{4.12}$$

where α is a scalar step length and g indicates the search direction of the objective function. In this book, the inverse problem is carried out by the iterative linearized approach using the steepest-descent algorithm, which means that the direction of the gradient is the search direction $g^{(k)} = -G^{(k)}$. Moreover, the step length can be obtained to minimize the objective function along the search direction [247].

4.3 NUMERICAL MODELING

4.3.1 ELASTIC MODEL BASED ON FE METHOD

FE simulations were carried out on a 10 mm thick aluminum plate with the size of 1100×1100 mm^2 and the material properties are given in Table 4.1. Cubic shape eight-node elements with a size of 1 mm were used in the mesh which ensured that more than 30 elements per wavelength at the highest frequency were exploited for accurate modeling [98]. The corrosion patches with desired shapes were modeled by removing the elements from the mesh. The plate was surrounded by an absorbing region to avoid reflections coming from the edges [33]. The waves were monitored by a circular array with a diameter of 700 mm consisting of 64 nodes acting as transducers as shown in Figure 4.3. Out-of-plane displacement was applied in one of the nodes and the wave-fields were measured by the other 63 nodes. This was repeated for all the source-receive combinations resulting in a 64×63 signal matrix. The source excitation used in simulations was a 5-cycle Hann-windowed tone-burst signal at a central frequency of 50 kHz, which had around 15 dB bandwidth from 35 to 65 kHz. By using the normal force at this frequency, it ensures that nearly pure A_0 mode is generated [17]. Also, around this frequency, the A_0 mode is highly dispersive and thus sensitive to thickness change, as shown in Figure 1.2(a). More details about the mode selection for guided wave tomography have been discussed by Huthwaite [94].

4.3.2 ACOUSTIC MODEL BASED ON FD METHOD

FD simulations were performed in a 2D space domain without considering the thickness of the plate. Under this approximation, the 2D acoustic model approximates the

Table 4.1
Material properties [169]

Material	Young's modulus (GPa)	Poisson's ratio	Density (kg/m^3)
Aluminum	70	0.33	2700
Steel	210	0.287	7850

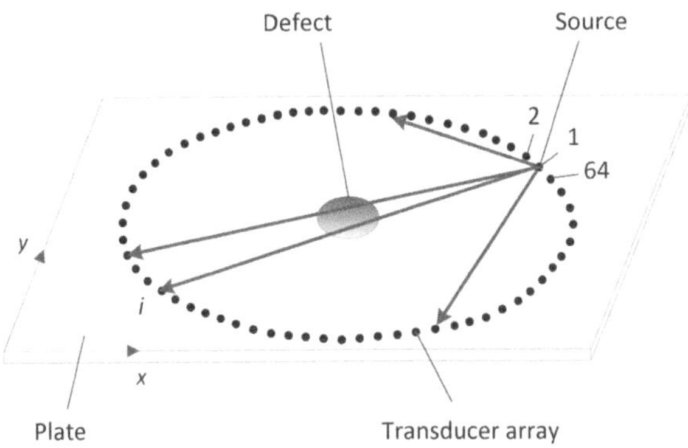

Figure 4.3 The configuration of a circular transducer array for UGWT on a plate [170].

behavior of dispersive guided waves by using the equivalent velocities in the model. The thickness variations in the plate are described by the respective velocity variations calculated from the dispersion relations characterizing the guided wave in the model. In this model, mode conversions do not occur. This is consistent with the 3D FE model, where the thickness variations of the defect are smooth, thus causing fewer mode conversions [42]. However, in the case of sharp discontinuities in the waveguide, the mismatch between the two models increases which could result in larger thickness reconstruction errors. A plate with the same dimension as in FE models was meshed by grid points with 2 mm spacing. This guarantees the calculation accuracy which requires at least 4 grid points per shortest wavelength [87]. The plate was surrounded by absorbing areas with the same width as in FE models. Omni-directional pressure waves were excited and monitored with a similar array configuration as described in the FE setup.

4.3.3 DATA PROCESSING

As the inversion algorithm is based on the FD acoustic modeling, the input scattered signals from the FE elastic modeling should be calibrated to account for possible deviations between the two models. One way to reduce the amount of phase and amplitude errors of the input is to match wave propagation results of two homogeneous models (without scatterers) by introducing the calibration factor defined as [97]

$$Q = \frac{fft^*(B_{FD})}{fft^*(B_{FE})}, \tag{4.13}$$

where fft represents the fast Fourier transform, $*$ is the complex conjugate, and B_{FD} and B_{FE} are the data from receivers in homogeneous models by FD and FE methods, respectively.

Additionally, a Gaussian filter is used to smooth the background before subsequent iterations. The aim of this filter is to minimize the effects of artifacts from each iteration. The correlation length of the Gaussian filter selected is around $\lambda/2$.

After the inversion, the reconstructed velocity maps from the FD and FE models are converted to thickness maps using the known dispersion relationship between phase velocity, thickness, and frequency.

4.3.4 FREQUENCY SELECTION AND CONVERGENCE

The iteration over sequential frequencies enables the inversion to start from a homogeneous velocity model and move toward the desired high-resolution true velocity model. At each frequency, the iterated model should provide the updated model with a close neighborhood to the global minimum for the next frequency. Global minima can be found more easily at lower frequencies where the velocity model resembles the homogeneous background model [29]. However, the well-defined strategy for selecting appropriate frequencies in waveform tomography has not yet been established [231]. Here we propose the largest frequency step between the frequency iterations according to the wavenumber k distribution of the scattered wave field from diffraction tomography [219]. The scattered field contains the transmission part which is limited within the circle of radius $\sqrt{2}k_1$ in the K space and the reflection subset which is contained between $\sqrt{2}k_1$ and $2k_1$ [18], as shown in Figure 2.1. Transmission wavenumbers at the lower frequency f_1 are always included in the larger wavenumber subset at the frequency f_2 within the circle of radius $\sqrt{2}k_2$. In order to account for all the reflection wavenumbers in the inversion, the relationship between two subsequent modelings at wavenumbers k_1 and k_2 should satisfy the relation $k_2 \leq \sqrt{2}k_1$. Thus for the inversion based on non-dispersive waves, the relationship between the frequencies is direct $f_2 \leq \sqrt{2}f_1$. This is slightly a stricter criterion for the current study as the wave propagation is dispersive and the growing rate of wavenumber in frequency is slower than non-dispersive waves.

The convergence of the results over iterations is examined by comparing the average relative thickness change E around the thinnest area with respect to the nominal thickness, similarly as in [97],

$$E_j^{(i)} = \frac{\overline{|T_j^{(i)}(x,y) - T_j^{(i-1)}(x,y)|}}{T_0}, \tag{4.14}$$

where $T_j^{(i)}(x,y)$ is the thickness at location (x,y) for the current iteration i at jth frequency component, and T_0 is the nominal thickness. The value of $E \leq 10^{-2}$ is accurate for the thickness estimation in engineering applications. However, the threshold for stopping the inversion is set to be $E \leq 0.5 \times 10^{-3}$ which is a stricter limit due to the slow conversion rate of the FWI algorithm. It should be noted that at lower frequencies a fixed number of iterations are carried out and the value E is approaching the convergence, while the above convergence criterion is only checked at the highest frequency. There are possibilities to optimize the choice of frequencies and the number of iterations, but this is beyond the scope of the current study.

4.4 EXPERIMENTS

The experimental setup is presented in Figure 4.4(a). Experimental measurements were performed on two 1100 mm×1100 mm×10 mm aluminum plates. A flat-bottomed circular hole with a wall boundary having an angle of 30° to the plate surface was machined in the first plate. This hole has a diameter of 60 mm on the plate's surface and a depth equivalent to 50% of the plate's thickness. The second plate contains an irregularly shaped defect with a maximum depth of 5 mm, which was produced by a computer numerical control (CNC) milling machine, and the defect was constructed by removing the material layer by layer. A zoomed picture of the defect is shown in Figure 4.4(b).

Measurements were taken along a circle with a 700 mm diameter, centered around the defect. The circle was marked with 64 generator/monitor positions, evenly

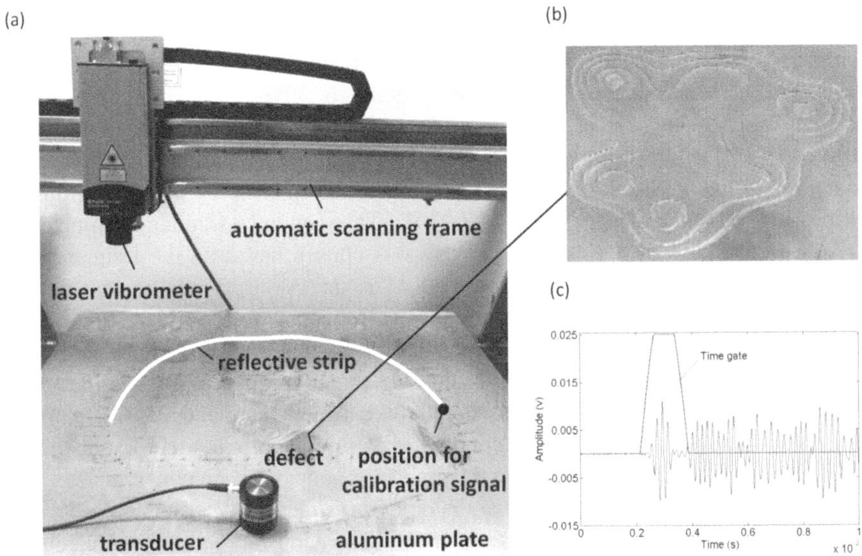

Figure 4.4 (a) The experimental setup featuring a complex-shaped defect, (b) a zoomed picture of the defect, and (c) a representative time trace from the experiment, indicating the selected time window [170].

spaced. A PZT transducer (Panametrics V1011) generated the A_0 guided mode at one position, while a Polytec OFV-505 laser vibrometer took measurements at the remaining positions. During each measurement, a 5-cycle Hann-windowed tone-burst signal at 50 kHz was generated by a Tiepie Handyscope HS3. Figure 4.4(c) shows a typical signal captured by the vibrometer. A gating function, akin to the one described in [97], was applied to remove unwanted components and obtain the first arrival wavepacket. To prevent edge reflection interference, only transmitted signals were used, resulting in measurements from half of the circle (33 signals), as depicted in Figure 4.4(a). This process was repeated 64 times to assemble a matrix of 64×33 signals.

Similar to the simulations, the measured signals need to be calibrated using Equation (4.13) before they can be used for inversion. However, in the experiment, due to the presence of the defect, signals in a homogeneous medium can only be obtained from limited positions. As a result, we measured the waveform at only one receiver position (as shown in Figure 4.4(a)), which was not affected by the scatterer, and then mapped this to all other receivers based on their positions. Given the limited anisotropy in sound speed in this plate, this approach is considered reasonable.

4.5 RECONSTRUCTION RESULTS

4.5.1 SINGLE REGULAR DEFECT

The first modeling involved a circular defect positioned at the center of a 64-element circular array, as depicted in Figure 4.5(a). The defect is a flat-bottomed circular hole with a stepped wall boundary. It has a surface diameter of 60 mm ($\simeq 1.3\lambda$ at 35 kHz, 1.8λ at 60 kHz) and exhibits a thickness reduction of 50%.

Figures 4.5(b), (c), and (d) present the monochromatic reconstruction of the thickness at 35 kHz after 40 iterations for the FD model, the FE model, and the experiment, respectively. Similarly, Figures 4.5(e), (f), and (g) display the polychromatic reconstruction of the plate thickness at 60 kHz. The sequential frequency group comprising frequencies of 35, 46, and 60 kHz was chosen based on the frequency selection criterion discussed in Section 4.3.4. The inversion process began with the homogeneous background model at the lowest frequency, followed by 20 iterations at each frequency. The velocity map obtained from the final iteration at each frequency served as the starting model for the subsequent frequency. The computations were performed on an HP Z820 workstation with a single core and 128 GB of memory. Each forward model in the FD simulation required approximately 25 seconds to solve, while the inversion process took around 1.5 hours to reach convergence.

It can be seen from the figures that the defect was reconstructed in all cases, with sharper images observed at higher frequencies. The acoustic modeling yielded very clear images. In the elastic model, some artifacts were visible around the array location, but they did not significantly affect the quality of the interior of the resolvable area. These artifacts are attributed to the under-sampling of the wavefield by the circular array. According to Simonetti et al. [218], the minimum number of transducers in a circular array required to adequately sample the wavefield can be expressed as

$$N > \frac{4\pi r_i}{\lambda}, \tag{4.15}$$

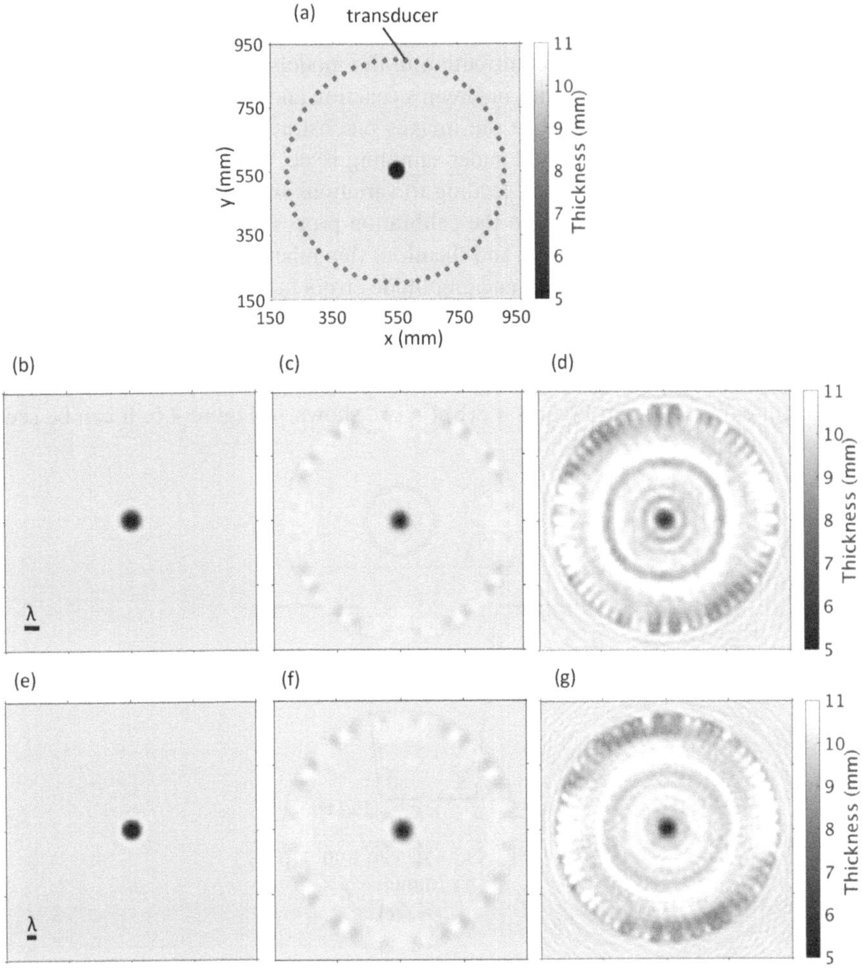

Figure 4.5 FWI reconstructions with a single central defect: (a) original model; (b) monochromatic 35 kHz in the acoustic model, (c) elastic model, and (d) experimental data; (e) polychromatic 60 kHz in the acoustic model, (f) elastic model, and (g) experimental data [170].

where r_i is the radius of the image to be free from grating lobes. At 60 kHz ($\lambda = 33.4$ mm), approximately 130 transducers would be required to accurately reconstruct an image for a 700 mm diameter area. However, it is not practical to have too many transducers in the experiment and therefore the number of transducers is limited to 64 in this study with the expected reduction in the imaging area that is free from grating lobes. It should be noted that in many field applications, the placement of transducers is constrained to a limited range of angles, further limiting the available information. Another possible reason for the artifact at the array position could be the inability to completely cancel the source contribution in the models during the inversion. This issue arises from the singularities in Green's function and its near-field effects [163]. Notable artifacts are observed in the images reconstructed from the experimental data. Despite the aforementioned under-sampling issue, the excited wavefields may not be perfectly omnidirectional, leading to variations in amplitude and phase with respect to the excitation angle. In the calibration process, signals without the scatterer are synthetically generated, and therefore this mismatch between the true and synthesized signal leads to phase and amplitude errors in the inversion. Additionally, the results can be affected by noise, measurement positioning errors, the length of the gating function, and the slight anisotropy of the material [16].

Comparisons between the reconstructions and the original thickness profile across the defect extracted from Figures 4.5(b)-(g) are shown in Figure 4.6. It can be seen

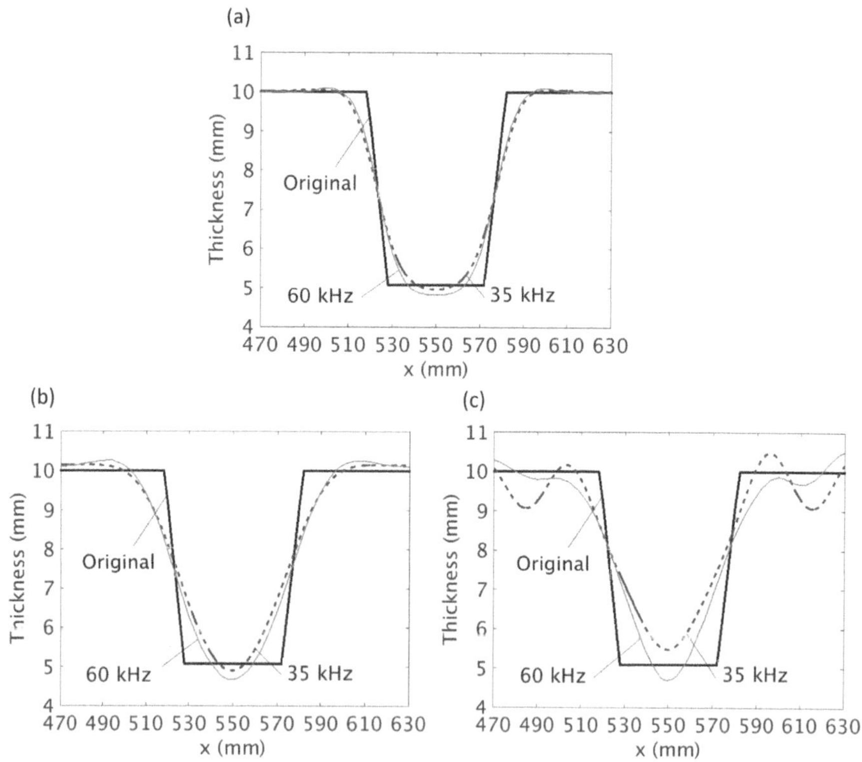

Figure 4.6 Cross sections of thickness reconstructions of a single central defect along the central line in (a) acoustic model, (b) elastic model, and (c) experimental data [170].

that in both FD and FE cases, the depth of the hole is already accurately reconstructed at 35 kHz. However, the reconstruction from experimental data slightly underestimates the depth by 0.4 mm. At 60 kHz, the reconstruction in all cases slightly overestimates the depth, with the largest error obtained from the experimental data being 0.3 mm from the true value. It is worth noting that the thickness reconstruction from the experiment using only the transmission part (a matrix of 64×33 signals) is similar to the simulation results based on both reflection and transmission parts (a matrix of 64×63 signals). These results clearly demonstrate a significant improvement in resolution compared to the resolution of TT, i.e., $\sqrt{L\lambda}$ = 153 mm (at 60 kHz).

Note that an attempt was made to directly perform the inversion at 60 kHz using a homogeneous velocity model. However, convergence was not achieved, indicating that the starting model was not accurate enough for the inversion. This is because at lower frequencies, the initial estimate captures longer-scale components with slow-varying features and fewer minima [29]. Consequently, the neighborhood of the global minimum is more likely to be captured at such a scale.

4.5.2 IRREGULARLY SHAPED DEFECT

The reconstruction of the irregularly shaped defect, as observed in the experiment, is shown in Figure 4.7(a). The depth of the defect varies randomly up to 5 mm, and its largest extent measures around 220 mm. The defect is characterized by complicated shape and smooth variations in thickness. Three sequential frequencies 35, 46, and 60 kHz were used for the inversion, and at each frequency, 30 iterations were applied. The forward model in this case took 50 seconds to solve, slightly longer than the previous case due to the complexity of the velocity map. The inversion process took around 4.5 hours to produce the results presented in Figures 4.7(b), (c), and (d).

It can be seen that the original shape of the defect, with its fine details, is reconstructed very well in the acoustic model. The reconstruction based on the elastic model can also accurately capture almost all the details of the defect. In the image reconstructed from the experimental data, there is more noise, as explained previously. However, the overall shape of the defect and its thinnest parts match the original image very closely.

The cross-sections of the largest corrosion depths in Figure 4.7 are depicted in Figures 4.8(a) and (b). It can be seen that the profiles of the defect are very accurately reconstructed using the data from the acoustic model. The reconstructions based on the data from the elastic model and the experiment are also very good, although the deepest points in the defect are slightly underestimated, with an error of approximately 0.5 mm. Notably, the reconstruction results from the experimental data and the elastic model using FE simulations are very close to each other, indicating that the inversion can be reliably performed using only the transmitted signals. This observation is consistent with previous work by other researchers [18].

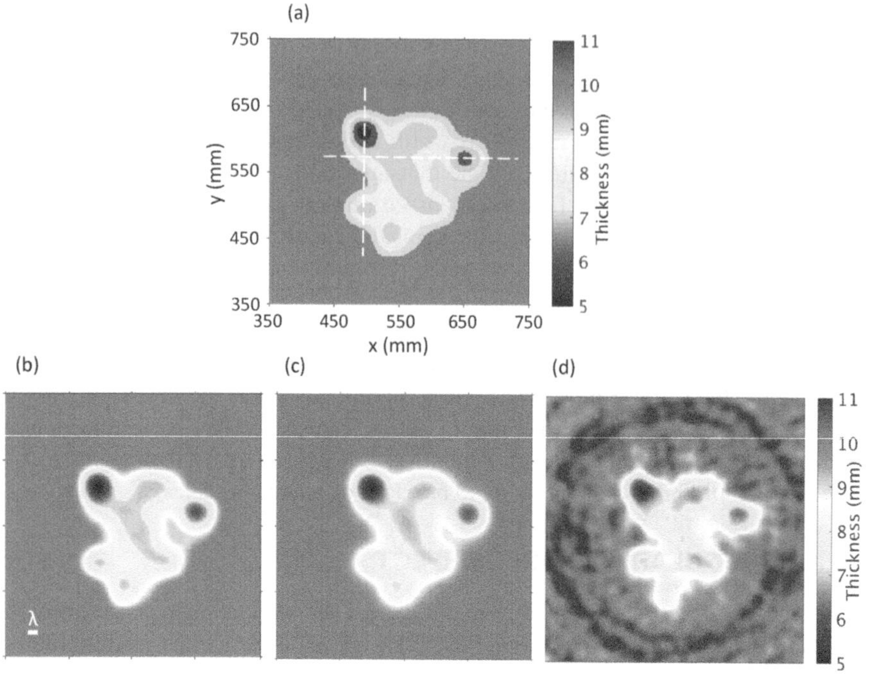

Figure 4.7 Polychromatic FWI reconstructions applied to irregular defects at 60 kHz. (a) Original model, (b) acoustic model, (c) elastic model, and (d) experimental data [170].

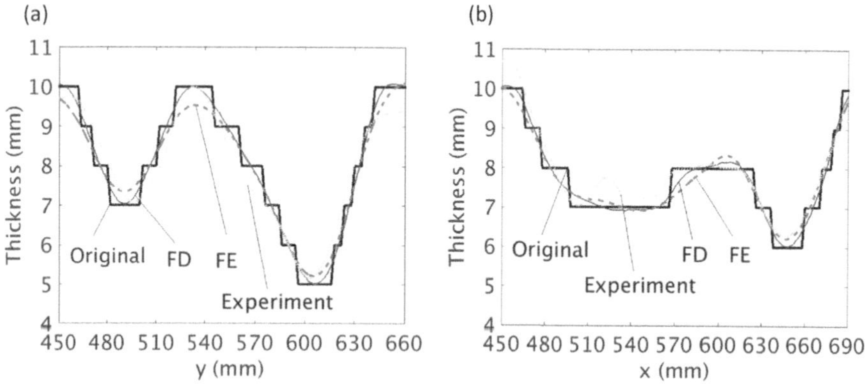

Figure 4.8 Cross sections of the thickness reconstruction of a complex defect along the lines shown in Figure 4.7 along (a) the vertical line and (b) the horizontal line [170].

4.6 DISCUSSION

The results presented in this chapter for two different defects have demonstrated that FWI is a valuable tool for obtaining high-resolution thickness mapping of plate-like structures using only transmission measurements of the total field. However, the results of the irregular defect also indicate that the resolution of FWI is slightly compromised when defects were reconstructed from the elastic model and experimental data compared with the equivalent reconstruction by using the simplified acoustic model. This is due to the fact that the 2D FD model is an approximation of the realistic 3D wave propagation model and may impose certain limitations on the inversion if the modeling data significantly deviate from each other. This finding is consistent with the work of Huthwaite [90], where the scattering from small scatterers in the two models within the Born approximation was compared. It was observed that the scattered waves from the 2D acoustic and 3D elastic models exhibit similar behavior only within a limited scattering region in the transmission zone.

One of the benefits of FWI lies in its multiple-frequency strategy, which allows for the processing of data subsets to improve resolution and incorporate smaller wavenumbers into tomography models [249]. The inversion can be started from the homogeneous background model at low frequency, where the global minimum can be more easily found as the velocity errors in the waveforms remain below a half-cycle [163]. As mentioned earlier, TT provides accurate reconstructions when the object size to be imaged is much larger than λ and the size of the first Fresnel zone $\sqrt{L\lambda}$ [18]. For the DT to be applicable, the phase shift traveling through the defect must be small. FWI serves as an alternative approach to bridge the gap between traveltime tomography and diffraction tomography. The results of the irregular defect demonstrate the success of the inversion, even when the defect size exceeds the first Fresnel zone, and the smaller details of the defect, approximately the size of the wavelength, can still be determined.

Compared to existing methods, the FWI approach has a major limitation in terms of computational complexity and cost. The relationship between the model and the measurement data is nonlinear, requiring numerous iterations for convergence [249]. The calculation results showed that the convergence for a single frequency can be achieved in 20 to 40 iterations. Depending on the total number of frequencies needed for the inversion, much more iterations may be required. While the current computational cost is acceptable for offline imaging, there is potential to enhance the time efficiency of FWI.

In this work, the conventional steepest descent algorithm was employed for numerical optimization. The convergence rate could be significantly improved by using more advanced optimization methods, such as the quasi-Newton algorithm [27]. Another approach to reduce the number of iterations is to utilize a low-resolution tomography algorithm to construct a more accurate starting model, which can then be refined using FWI. Additionally, the sequential execution method employed to solve the forward problem in this work could be replaced by the massively parallel solver MUMPS algorithm [3]. This algorithm distributes the solving processes across multiple processors, significantly improving calculation speed by more than one order of magnitude.

4.7 SUMMARY

In this chapter, a mono-parameter FWI method is developed for UGWT on plate-like structures. The method consists of a forward solver that predicts the scattering measurements in a 2D acoustic model, and an inverse model that iteratively updates the velocity map, which is then connected to the thickness map through the dispersion relations of selected guided wave modes. The algorithm utilizes a multiple-frequency strategy, where the image generated at a lower frequency serves as the starting model for the higher frequency. This approach ensures that the reconstruction of the thickness map is less reliant on the initial model, while maintaining high resolution for small features.

The algorithm is demonstrated using both simulations and experiments with a simple circular hole and a complex-shaped defect. The results show that the minimum thickness can be reconstructed with an accuracy of within 1 mm for a 10 mm thick plate. The FWI method allows for the consideration of higher-order scattering effects in the model, which has the potential to enhance resolution compared to other guided wave tomography methods. Further investigation of these details will be presented in the next chapter.

4.8 EXERCISES

1. Please write the time domain and frequency domain 2D acoustic wave equations. Try to analyze the advantages and disadvantages of time domain and frequency domain full waveform inversion.
2. Please try to analyze the advantages and disadvantages of 3D full waveform inversion compared with 2D full waveform inversion used in this chapter.
3. How is the mesh size selected in finite difference and finite element method analysis?
4. In addition to using the steepest descent method mentioned in this chapter, what other methods can replace the steepest descent method?
5. How to choose the appropriate multiple frequencies in frequency domain full waveform inversion?
6. How is selected the spacing between the two transducers mentioned in this chapter? Please explain.
7. In the two examples mentioned in this chapter, why is the error of inversion of experimental data larger than that of simulation data?
8. In the inversion model, how to choose the number of iterations?
9. How to determine whether it is converged?
10. In data processing, how to do calibration? Are there other calibration methods?
11. How can the calculation time of the current model be reduced?
12. The transducer array used in this chapter is circular. Can it be designed into other shapes? Like a rectangular array? Linear arrays?
13. What is a tone-burst signal?

14. Can the full waveform inversion be applied to other materials and structures, such as defect detection of anisotropic materials or multilayer structure materials?
15. Please plot the Lamb wave dispersion curves, including phase velocity and group velocity, in the aluminum plate.
16. Please analyze whether it is possible to use the L1 norm in addition to choosing L2 in solving the data residual (see Equation (4.4)).
17. Are acoustic models and elastic models for forward model or inversion model?
18. Please explain the difference between the finite difference method and the finite element method.

5 Reconstruction Accuracy

5.1 INTRODUCTION

A number of studies have been carried out to investigate UGWT algorithms, such as TT [131], DT [18, 147], the HARBUT algorithm [96], and the mono-parameter FWI approach (as discussed in Chapter 4). One of the primary concerns in UGWT is the resolution and accuracy of the reconstruction, prompting research efforts aimed at evaluating the performance of various algorithms.

TT, based on ray assumptions, is typically effective in reconstructing large and smoothly varying defects, with the minimum resolvable size often determined by the first Fresnel zone. DT offers improved resolution by utilizing a linearized scattering model. However, DT's theoretical resolution limit of $\lambda/2$ is often compromised by the neglect of multiple scattering effects, resulting in the reconstruction of only small or low-contrast defects. The HARBUT algorithm has been developed to address the reconstruction of complex-shaped defects. The theoretical resolution limit of HARBUT is the same as that of DT. While its theoretical resolution limit matches that of DT, the achievable resolution based on the elastic model is reduced to around 1.5-2λ due to the simplifications in the acoustic inversion model for guided wave propagation. To further enhance accuracy, an approach based on Mindlin plate theory has been developed to incorporate scattering from guided waves into the inversion process [199]. However, this model also relies on the Born approximation and assumes small perturbations in local properties, such as small thickness loss, which limits its applicability to a specific range of defects. The development of UGWT with an improved scattering model allows for a more accurate representation of scattering behavior and significant improvement in reconstruction accuracy [91].

This chapter provides a comprehensive discussion on the reconstruction accuracy of the mono-parameter FWI approach. In comparison to DT, FWI has the advantage of considering higher-order diffraction and scattering effects in its numerical solver, which enables more accurate inversion results, particularly in the presence of multiple defects. The accuracy of the reconstruction is also influenced by the shapes of the defects, and therefore defects of various shapes have been studied. The modeling and experimental procedures are outlined in Section 5.2, followed by an analysis of FWI's performance for both single and multiple defects in Section 5.3. The results from multiple defects are validated through experimental measurements and compared with the results obtained from DT in Section 5.4. The chapter concludes with a summary of the findings in Section 5.5.

DOI: 10.1201/9781003667926-5

5.2 VALIDATION METHODS

5.2.1 DEFECTS

The configuration is presented in Figure 5.1. A circular array of 64 transducers surrounds the defect. Two shapes of artificial defects were considered in this chapter. One of them has an axisymmetric Hann-shaped profile, as shown in Figure 5.1(a). This shape can be described mathematically by

$$T(r_d) = \begin{cases} T_0 - \frac{D_d}{2}\left[1+\cos\left(\frac{2\pi r_d}{W}\right)\right] & r_d < \frac{W}{2} \\ T_0 & r_d \geq \frac{W}{2}, \end{cases} \tag{5.1}$$

where T_0 denotes the nominal thickness, and r_d is the radius of the defect at a specific depth. This profile exhibits a smooth variation in thickness, characterized by the deepest depth D_d and the surface diameter W. The smooth change in thickness provides valuable insights into the performance and the limitations of the tomographic algorithm itself, as it helps suppress the complexity of actual corrosion patches. Additionally, no mode conversions are expected from smooth thickness changes that cannot be modeled in the acoustic model. While these defects may not fully represent real-world scenarios, complex defects often exhibit similar behaviors as the superposition of simpler defects [90,93]. Therefore, trends identified for simpler defects can be expected to manifest in complex defects as well.

The second defect type is the flat-bottom defect with a variable wall angle. It is defined by the surface diameter W_1, the bottom diameter W_2 and the defect depth D_d, as depicted in Figure 5.1(b). This model is useful for describing sudden changes in thickness, where mode conversions could occur, therefore impacting the inversion performance [17, 37, 116]. Additionally, three flat-bottom defects of varying sizes and depths were modeled in the same plate, simulating a more complex scenario involving a closely-positioned defect cluster. Here, multiple scattering effects might arise, potentially affecting the FWI reconstructions. Defect A was placed at the center of the plate, with W_1 = 60 mm, W_2 = 44 mm, and D_d = 5 mm. Two smaller flat-bottom circular defects B and C were located 100 mm from defect A in the vertical and horizontal directions, respectively. Defect B had dimensions W_1 = 42 mm, W_2 = 30 mm, and D_d = 4 mm, whereas the smallest defect C had W_1 = 24 mm, W_2 = 16 mm, and D_d = 3 mm.

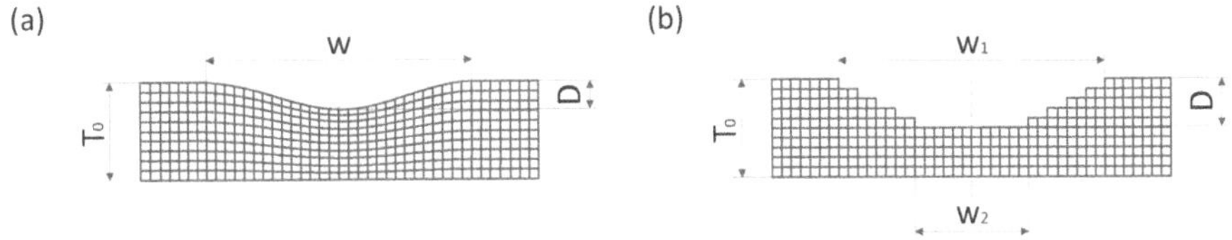

Figure 5.1 (a) Cross-section of a defect with a Hann shape, and (b) cross-section of a flat-bottom defect, both taken along the central line [172].

5.2.2 NUMERICAL MODELING

FE simulations were conducted on two different cases to analyze the resolution of FWI. The first case involved a series of single Hann-shaped defects of various sizes and depths located at the center of an aluminum plate (see Table 4.1), with dimensions of 400 mm×400 mm×10 mm. The cubic-shaped brick elements with an edge length of 0.5 mm were chosen, as they were deemed fine enough to accurately model this case [98]. The absorbing region was introduced to all sides of the plate to eliminate reflections from the edges [168]. An array of 64 transducers with a diameter of 360 mm was modeled, as shown in Figure 5.1. For a given excitation point, the A_0 mode was generated by using an out-of-plane force, and out-of-plane displacements were measured at the other 63 receivers. By repeating this process, a signal matrix of 64×63 can be recorded by combining all source-receive pairs. The input signal was a 2-cycle Hanning-windowed tone-burst signal centered at 70 kHz, with a bandwidth of approximately 23 dB ranging from 10 to 130 kHz. In this frequency range, the phase velocity of the A_0 mode depends on the frequency and, therefore, could be sensitive to variations in the plate's thickness.

The second case considered three flat-bottom defects located at the center of an aluminum plate with the size of 1100 mm×1100 mm×10 mm. For this simulation, a circular array of 64 transducers with a diameter of 700 mm was used. The edge length of the cubic-shaped, eight-node element was set to 1 mm. A 5-cycle Hanning-windowed tone-burst signal centered at 50 kHz, featuring a bandwidth of approximately 15 dB between 35 and 65 kHz, was employed. The signals from the simulations were collected in a manner akin to the model with a single Hann-shaped defect.

The FD model was meshed with no fewer than 4 grid points per shortest wavelength, as detailed in [87]. The absorbing regions and array configurations were modeled in alignment with the corresponding FE models. Additional details are available in 4.3.

5.2.3 EXPERIMENTAL PROCEDURES

The experimental setup comprised an aluminum plate of 1100 mm × 1100 mm × 10 mm with the same set of three defects as was described in 5.2.1, with a 64 element array with a diameter of 700 mm, as shown in Figure 5.2. The PZT transducer was used to excite A_0 at one position, and a laser vibrometer was used to measure the displacement perpendicular to the surface. In each measurement, a 5-cycle Hanning-windowed tone-burst signal centered at 50 kHz was generated by a Tiepie Handyscope HS3.

Given the limited size of the plate and to prevent reflections from the edges, only transmitted signals were used, and therefore measurements were taken around half of the array, as shown in Figure 5.2. This process was repeated 64 times, generating a 64 × 33 matrix for all source-receiver combinations. A gating function was employed to isolate the first arrival wavepacket from unwanted components in the measured signals. Prior to using the signals for inversion, they were calibrated employing the same method described in Section 4.3.3.

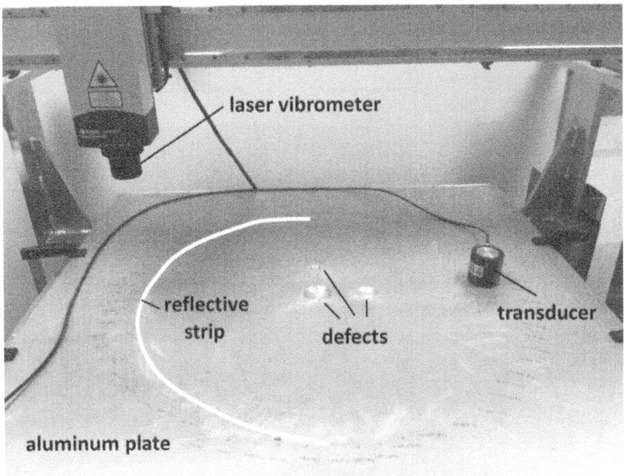

Figure 5.2 Experimental setup with three flat-bottom defects of various sizes and depths. The transducer was used to generate the guided wave, and measurements were taken by the laser vibrometer [172].

5.3 RESULTS

5.3.1 DEPTH ESTIMATION OF THE HANN-SHAPED DEFECT

The first case involves the reconstructions of the axisymmetric Hann-shaped defects at the center of the circular array, using a similar approach as demonstrated in [90]. Various defects were simulated, with depths of 1, 3, and 6 mm, and surface diameters of 30, 60, and 120 mm, corresponding to 0.9λ, 1.8λ, and 3.6λ at 60 kHz. Figure 5.3 displays the cross-sectional reconstructions from both the acoustic model and the elastic model at 60 kHz. A homogeneous background at the lowest frequency of 10 kHz is set as the initial model, with 40 iterations per frequency.

It can be seen that the reconstructions of the defects based on the acoustic model are highly accurate. The most significant error, about 0.4 mm from the true thickness, is observed for the narrowest and deepest defect, as shown in Figure 5.3(g). On the other hand, larger reconstruction errors can be observed for defects simulated in the elastic model. The smallest error in thickness, around 0.2 mm, occurs for the large and shallow defect illustrated in Figure 5.3(c). As the defects become smaller and deeper, the error progressively increases. The reconstruction in Figure 5.3(g) is notably unsuccessful. This lack of resolution may be due to the varying scattering behavior between guided waves with different thicknesses and acoustic waves with differing velocities. The 2D approximation utilized in the inversion contributes to inconsistency with the signals measured from the 3D elastic model [90].

In industries, one of the key interests in evaluating the residual life of a structure is determining the minimum thickness in the corrosion profile. Therefore, in UGWT, a quality parameter e_{depth} is used to describe the maximum depth error. This parameter

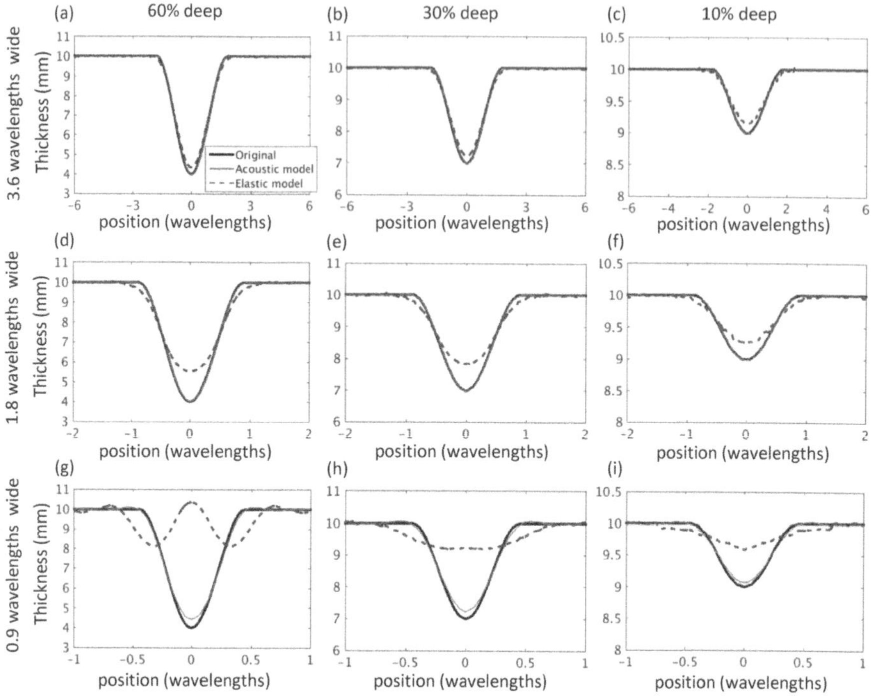

Figure 5.3 Cross sections of the thickness reconstructions of Hann-shaped defects along the central line from both acoustic and elastic modeling data at 60 kHz. The first row (a)–(c) has a surface width of 120 mm, the middle (d)–(f) has a width of 60 mm, and the final (g)–(i) has a width of 30 mm, The left column (a), (d), and (g) is 60% deep; the middle (b), (e), and (h) is 30% deep, and the right (c), (f), and (i) is 10% deep [172].

can be defined as follows [90]

$$e_{depth} = \frac{T_{min} - T_{min}^{res}}{T_0}, \tag{5.2}$$

where T_{min} and T_{min}^{res} represent the minimum thickness of the original and reconstructed thickness maps, respectively, while T_0 denotes the nominal thickness. Note that the maximum depth error is calculated using the nominal thickness of the plate, not the defect depth, resulting in smaller errors for lower contrast defects. Figure 5.4(a) displays the maximum depth error in the reconstruction from the elastic model across various frequencies and depths for a surface diameter of 60 mm. The maximum depth error decreases as the frequency increases but grows as the defect depth increases. With a fixed depth, the accuracy of the reconstruction improves with increasing frequency. This improvement is primarily attributed to enhanced resolution at higher frequencies. Figure 5.4(b) presents the cross-sectional reconstructions of

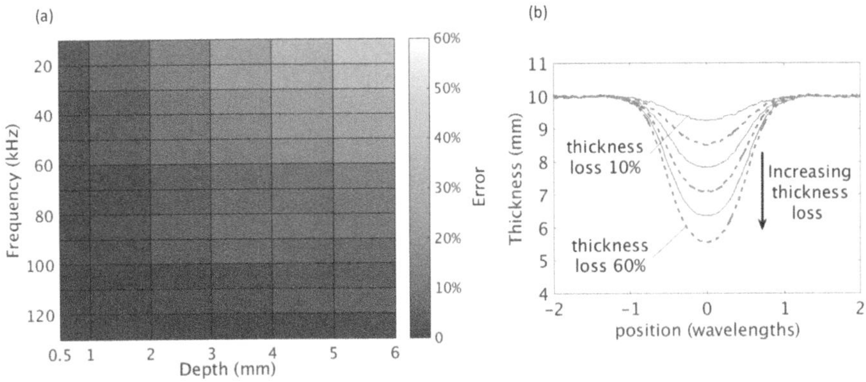

Figure 5.4 (a) The maximum depth error of reconstruction from the elastic model as a percentage of the nominal thickness. Errors are given for Hann-shaped defects with a surface diameter of 60 mm. (b) Cross sections along the central line of the reconstructed defects at various depths at 60 kHz, as derived from (a) [172].

the defect with a surface diameter of 60 mm at a single frequency of 60 kHz. The plot illustrates that the reconstructed depth increases as the true depth of the defect increases. However, it is important to note that when the thickness loss is employed to represent the reconstruction and is divided by the true defect depth, the curves nearly coincide, as previously demonstrated in [93]. Thus, it becomes clear that the maximum depth error is proportional to the defect depth.

5.3.2 IMAGING OF MULTIPLE DEFECTS

The second case investigates multiple defects, where three flat-bottom holes of varying sizes and depths are located closely on a plate, as shown in Figure 5.5(a). Defect A has a surface diameter of 60 mm ($\simeq 1.3\lambda$ at 35 kHz, $\simeq 1.5\lambda$ at 46 kHz, and $\simeq 1.8\lambda$ at 60 kHz) and the depth that is 50% of the thickness. Defect B has a surface diameter of 42 mm ($\simeq 0.89\lambda$ at 35 kHz, $\simeq 1.1\lambda$ at 46 kHz, and $\simeq 1.3\lambda$ at 60 kHz) and has a depth that is 40% of the thickness. Defect C, with a surface diameter of 24 mm, corresponds to approximately 0.51λ at 35 kHz, 0.6λ at 45 kHz, and 0.72λ at 60 kHz, with a thickness reduction of 30%.

Figures 5.5(b), (c), and (d) present the polychromatic reconstructed thickness mapping from the acoustic model, elastic model, and experiment at 60 kHz, respectively. It is evident that high-quality images for all three defects can be obtained using the acoustic model. The larger defects A and B are also well reconstructed in the elastic modeling, but the reconstructed image of the smallest defect C appears fainter and has a seemingly larger diameter than the original size. The images of defects A and B based on the experiment also align well with the true thickness mapping; however, the smallest defect C is barely visible. This suggests that the resolution is

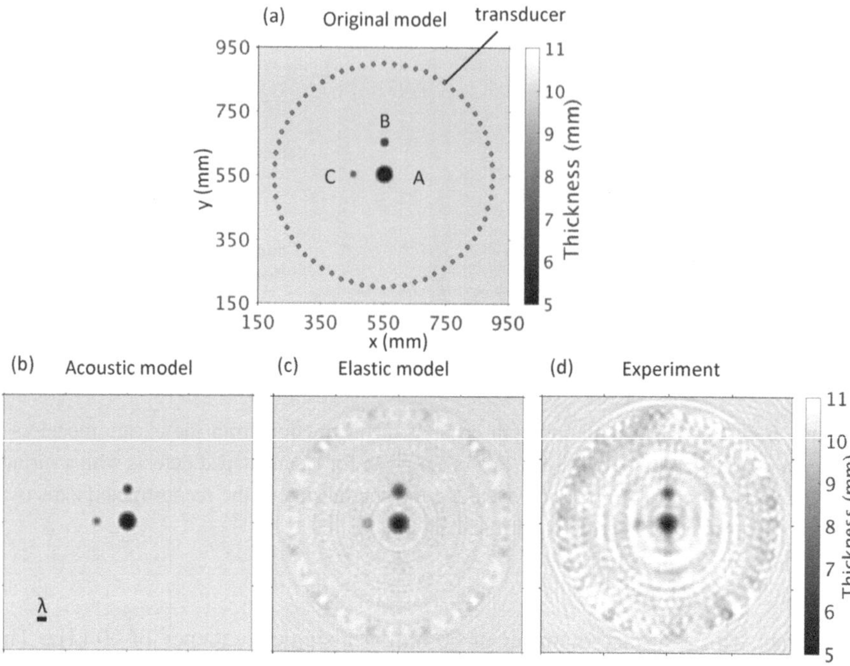

Figure 5.5 FWI reconstructions of three flat-bottom defects at 60 kHz. (a) Original model, (b) the acoustic model, (c) the elastic model, and (d) the experiment [172].

compromised when performing the inversion with elastic and experimental data compared to acoustic data. In the elastic model, some artifacts are observed near the position of the array. Additionally, the reconstructed image from the experimental data presents notable artifacts. The underlying reasons for these artifacts are discussed in Section 4.5.1.

Figures 5.6(a)-(f) show comparisons of thickness reconstructions along two cross-sections across defects A, B, and C, A for different frequencies at 35, 46, and 60 kHz. It can be seen that all defects are very accurately reconstructed in the acoustic model: defect A at 35 kHz ($\simeq 1.3\lambda$), defect B at 46 kHz ($\simeq 1.1\lambda$), and the smallest defect C at 60 kHz ($\simeq 0.72\lambda$).

However, the reconstruction error increases when using realistic data from the elastic model and the experiment, compared to the data from the acoustic model. The depth of defect A is underestimated at 35 kHz in the experiment and is slightly overestimated at 60 kHz in both the elastic model and the experiment. Defect B is reconstructed at 46 kHz (approximately 1.1λ) in the elastic model and at 60 kHz (approximately 1.3λ) in the experiment, with a depth error of less than 1 mm in both cases. Finally, it can be seen that neither the elastic model nor the experiment can reconstruct defect C, resulting in a depth error of around 2 mm. These results

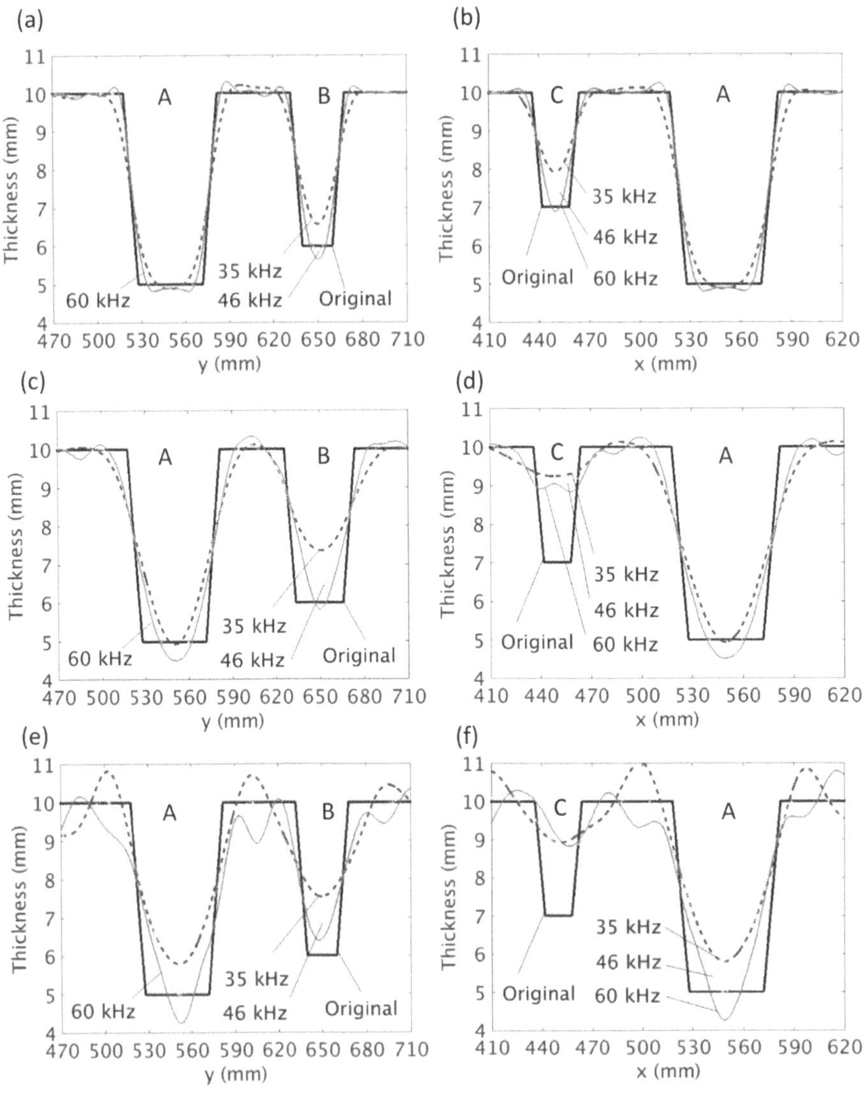

Figure 5.6 (a) Cross sections of thickness reconstructions of three flat-bottom defects at different frequencies based on FWI. The acoustic model along the center line of A and B, and (b) along the center line of C and A; (c) the elastic model along the center line of A and B, and (d) along the center line of C and A; and (e) the experiment along the center line of A and B, and (f) along the center line of C and A [172].

suggest that the resolution obtained in the elastic model and experiment is inferior compared to that of the acoustic model. The possible reasons for this discrepancy will be discussed in the following section.

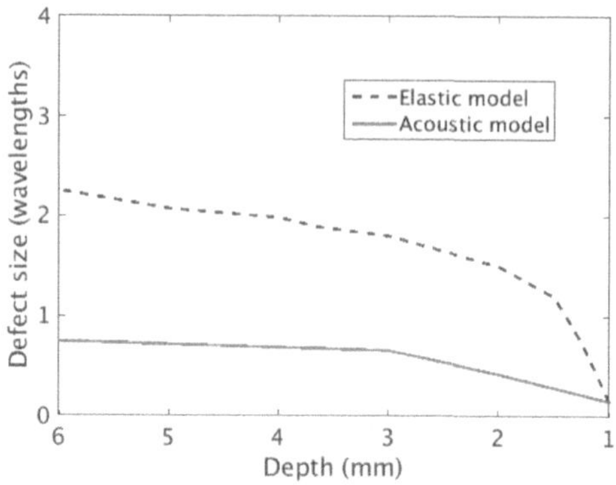

Figure 5.7 The acceptable resolution limit for different defect depths and sizes using the acoustic model and the elastic model at 60 kHz [172].

5.4 ANALYSIS AND DISCUSSION

5.4.1 RESOLUTION ANALYSIS OF THE HANN-SHAPED DEFECT

The main interest in UGWT is to determine the maximum depth of the defect in a corrosion profile. In this chapter, the reconstruction of the defect is considered to be accurate when the error in the maximum depth is less than 10%, i.e., $|e_{depth}| < 10\%$. Hence, the resolution limit is defined here as the wavelength corresponding to the maximum wavenumber which is accurately reconstructed in the image [90].

The reconstructions of defects with different depths (0.5, 1, 2, 3, 4, 5, 6 mm) and surface widths (5, 10, 15, 20, 30, 40, 60, 80, 120, 160, 240 mm) were analyzed using both the acoustic model and the elastic model at 60 kHz. The resolvable defect sizes of the two models are shown in Figure 5.7. It can be clearly seen that defects can be accurately reconstructed when their sizes are around 0.7λ and their depths are around 3 to 6 mm. Smaller defects with shallower depths can also be reconstructed from the acoustic model. Compared to the resolution achieved from the acoustic modeling, the resolution obtained from the more realistic FE modeling data drops to around 1.5-2λ for the majority of defect depths. This result is similar to previous studies using other tomography algorithms [90]. This suggests that, regardless of the inversion approach, the resolution limit is mainly constrained by the mapping from the simplified 2D acoustic model to the realistic 3D elastic model.

For further analysis of the resolution from the realistic test data, the relationship between the resolution limit and frequency for different Hann-shaped defects with a depth of 3 mm is presented in Figure 5.8(a). It can be seen from the figure that the resolution improves with the increase of the frequency. This could be expected because the wavelength decreases as the frequency increases, and the resolution limit is usually proportional to the wavelength [103]. In Figure 5.8(b), the achievable resolution limit normalized by the wavelength shows a slight upward trend at higher frequencies in general, indicating decreasing accuracy of the algorithm at higher

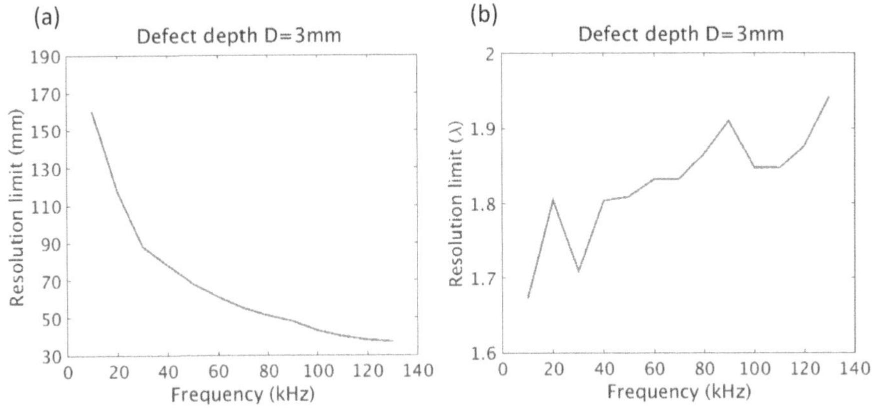

Figure 5.8 Resolution limit for the A_0 mode as a function of frequency obtained from the elastic model. (a) An absolute value, mm, and (b) relative to wavelengths λ for the 30% deep defect [172].

frequencies. The reason could be that the phase velocity of A_0 is approaching to plateau at higher frequencies, thus losing its sensitivity to the thickness. This suggests that a small error in the velocity can lead to a large error in the reconstructed thickness. Similar trends shown in Figure 5.8 were observed in other defect depths.

5.4.2 EFFECT OF THE BOTTOM WIDTH ON THE RECONSTRUCTION ACCURACY

In the previous section, the resolution limit was predicted for the defect with smooth thickness variation and a single deepest point. Results from Section 5.3.2 indicate that different levels of accuracy can be achieved for flat-bottom defects. It was found that the defects A and B were resolved at 1.3λ and 1.1λ, respectively, which are significantly better results compared with 1.5-2λ for the Hann-shaped defect from the realistic elastic data. It is clear that the shape of the defect profile affects the resolvability. Therefore, in this section, investigations are carried out to understand the influence on the reconstruction accuracy from the surface diameter W_1 and the bottom diameter W_2 of the flat-bottom defect. Figures 5.9(a) and (b) show the maximum depth errors as a function of the ratio of W_2/W_1 for depths of D_d = 6 mm and D_d = 3 mm, respectively. Three surface diameters of W_1 = 42, 60, and 75 mm have been selected, and the reconstructions have been performed at 60 kHz to obtain the defect sizes of around 1.25λ, 1.8λ, and 2.25λ. In general, trends of e_{depth} are similar for different surface diameters at the corresponding depth, while smaller errors were observed in the shallower defects, as shown in Figure 5.9(b). When the ratio of W_2/W_1 is small, the defect depth is underestimated, similarly as was shown in the reconstructions of Hann-shaped defects. For example, in Figure 5.3(d), the reconstruction of the defect with W = 60 mm and D_D = 6 mm at 60 kHz gives a depth

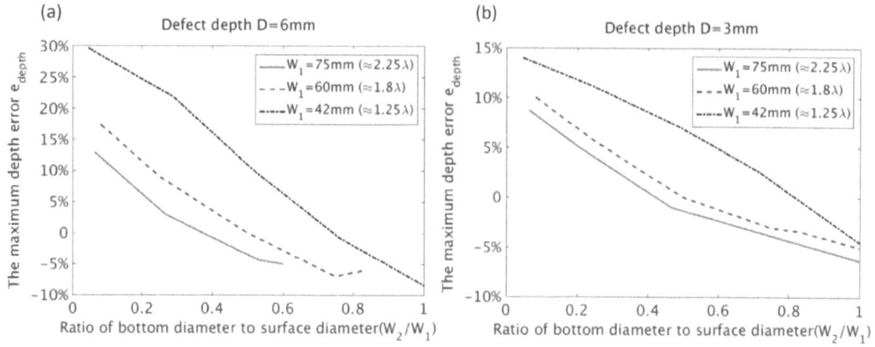

Figure 5.9 Maximum depth errors of flat-bottom defects at 60 kHz, using data from the elastic model, plotted against different ratios of bottom diameter W_2 to surface diameter W_1 for depths of (a) 6 mm, and (b) 3 mm [172].

error of 15%, which is the same as in Figure 5.9(a), when the ratio is 0.12. The error is caused by neglecting the presence of additional scattering terms related to density variation in the 2D inversion model [91, 92]. For a fixed surface diameter, it can be seen that the depth error decreases with the bottom diameter increasing. After the ratio is within a range of W_2/W_1 = 0.4-0.8, the error becomes negative, which means that the reconstruction overestimates the depth. This can be identified from defect A (ratio≃ 0.7), as shown in Figure 5.6(c). Finally, a considerable overestimation of the thickness at W_2/W_1 = 1 can be seen. Low-pass filtering is a feature of the image algorithm, which leads to a diffraction-limited resolution limit proportional to the wavelength in the reconstruction. All spatial frequency components above a certain value are removed, resulting in the overestimation of defect depth. Similar results have also been discussed in ref [93].

In summary, when the surface diameter approaches the resolution limit derived from the elastic data (approximately 1.8λ), the reconstruction error for defects with sharp thickness variation decreases relative to that for smoothly varying defects. Larger defects, characterized by a sharper change in thickness, tend to lead to a greater overestimation in depth compared to smaller defects, even when both have the same large ratio.

5.4.3 COMPARISON WITH DT ON THE IMAGING OF MULTIPLE DEFECTS

In practice, one important scenario is imaging multiple defects, which was investigated in Section 5.3.2. In this section, the same defects are reconstructed using DT, which is based on the Born approximation, and the results are compared with those obtained from FWI. Figure 5.10 displays the reconstruction of three flat-bottom defects using DT with elastic data at 60 kHz. Cross-sectional plots along the vertical

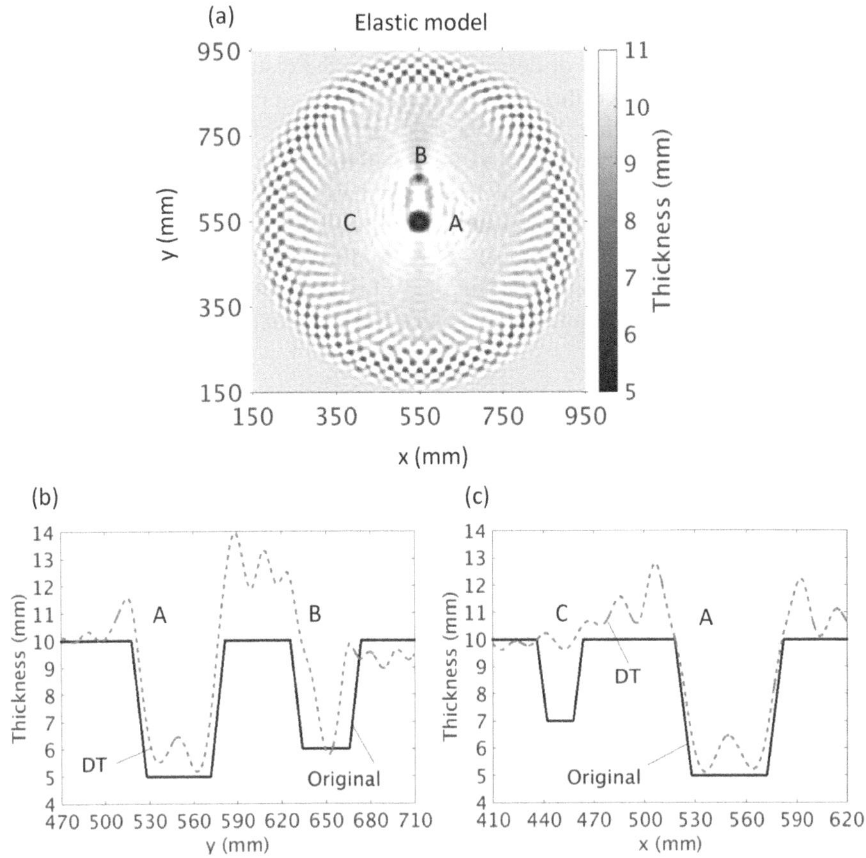

Figure 5.10 (a) DT reconstruction of three flat-bottom defects using the elastic model at 60 kHz, (b) cross sections of thickness reconstruction along the center line of A and B, and (c) cross sections along the center line of C and A [172].

and horizontal central lines are shown in Figures 5.10(b) and (c). When compared with the results using FWI in Figure 5.5(c) and Figures 5.6(c)-(d), it is clear that although defects A and B can still be identified, there are strong artifacts that completely obscure defect C. One of the reasons is a violation of the Born approximation, which stipulates that the phase shift between the incident field and the wave traveling through the defect should be less than π [221]. As the wave passes through several relatively large defects—each comparable in size to the wavelength—the accumulated phase delay can significantly exceed the ideal limit. However, FWI provides an advantage in this scenario. FWI uses a low-frequency, homogeneous background model during the inversion process, ensuring that the phase difference remains less than half a cycle [170]. Another challenge with DT is its neglect of higher-order scattering information in the inversion process, which can result in the form of coherent noise [217, 270]. The strength of the higher-order scattering depends on the size, contrast, and position of defects, as well as the distance between them. Previous research has shown that even when each component of a multi-component object is weakly scattering, the higher-order scattering effects become significant if the components align with the direction of illumination [13, 270]. In the current case, the second-order scattering fields are primarily generated from the blocking component, illuminating the component in the shadow area of the forward scattering wavefield. As such, the additional scatterer information encoded by higher-order scattering, which is taken into account in FWI, can be exploited to obtain a more accurate reconstruction than DT based on the Born approximation.

5.5 SUMMARY

In this chapter, the reconstruction accuracy analysis of UGWT based on mono-parameter FWI has been carried out to map the thickness of plate-like structures containing corrosion defects. Simulations of smoothly varying defects across varying widths, depths, and frequencies reveal that while an achievable resolution of 0.7 wavelengths can be obtained from acoustic modeling data, this drops to around 1.5-2 wavelengths when considering realistic elastic modeling data. This resolution limit aligns with previous studies using other guided wave tomography methods that employ velocity-to-thickness mapping approaches. The reduced accuracy can be attributed to the implementation of an approximate 2D inversion algorithm, which limits the complete utilization of data from the 3D elastic model. The analysis also demonstrates that the reconstruction accuracy of FWI is affected by the shape of the defect. Abrupt changes in wall thickness reduce the reconstruction error for small defects compared to those with smoothly varying thicknesses. However, larger defects with sharper changes in thickness are more likely to result in overestimations of depth. The imaging results of multiple defects, validated by experiments, demonstrated the good performance of FWI, as it allows the estimation and accounting of phase distortion by initiating reconstruction from low frequencies.

5.6 EXERCISES

1. What is the theoretical resolution of the full waveform inversion algorithm?
2. What is the reconstruction accuracy?
3. What are the factors that affect reconstruction accuracy?
4. Which two shapes of artificial defects are mentioned in this chapter?
5. What is the acceptable resolution limit for different defect depths and sizes of the acoustic model and the elastic model at 60 kHz? Please draw a schematic diagram. What is the reason for the gap between the two models?
6. Please describe the advantages and disadvantages of 2D full waveform inversion in terms of reconstruction accuracy.

7. Please compare the imaging results of multiple defects mentioned in this chapter, which are based on diffraction tomography and full waveform inversion. Please give reasons for your answers.
8. Please think about how to further improve the reconstruction accuracy of full waveform inversion. Please give one or two suggestions.
9. How does the bottom width of a flat-bottom defect affect the reconstruction accuracy?
10. How does the reconstruction accuracy of the Hann-shaped defects vary?

6 Limited View Tomography

6.1 INTRODUCTION

Ideal quantitative reconstruction using UGWT can be achieved when the measured data is collected from as many locations around the object as possible—a full view configuration. Thus far, this book has focused on full view problems. However, in many applications, only limited view angles are available. For instance, pipeline assessment using UGWT typically involves two transducer rings on either side of the inspection area, thereby restricting viewing angles. Reconstruction from incomplete data is ill-posed, indicating that more than one solution can fit the measured data [51]. The absence of certain angles typically results in noticeable artifacts, degrading the quality of reconstruction by altering the reconstructed velocities and generating ghost features [99]. Similar issues are widely prevalent in other related areas, such as X-ray tomography [8], seismology [55], and medical ultrasound applications [59]. Various efforts have been made to improve reconstruction from the limited view, which can be broadly classified into three categories: statistical probabilistic algorithms [114], minimization algorithms [112, 158], and projection regularization methods [99, 213].

Two representative approaches under statistical probabilistic algorithms are maximum entropy techniques [191] and Bayesian methods [80]. These methodologies facilitate the integration of prior information about maximum entropy and the general shape or structure of the object, respectively. When the prior information is sufficiently restrictive, accurate estimations of the unknown components can be obtained, thereby reducing image artifacts. The main challenge associated with these algorithms is their high computational complexity and sensitivity to errors in the data [56].

The minimization algorithm finds a solution that best fits the measured data while adhering to a specific constraint. Weighting variables are used to balance between the measured data and the constraint. In each iteration of the inverse model, the residual data (difference between computed results and the measured data) is minimized, converging ultimately to a single fitting minimum. This method is widely used for total variation minimization and is often combined with regularization [112, 158]. However, the performance of minimization algorithms largely depends on the selection of appropriate constraints, which can be complex. Moreover, these methods are also computationally intensive [53].

The projection regularization method can be combined with direct imaging approaches in reconstructions. This method involves a projection process where a regularization restriction is applied directly to the image at each step using specific constraints. A common example of this method is positivity, which imposes a positivity constraint on each pixel of the estimated image and sets negative values to zero [213]. Huthwaite [99] introduced a projection-based regularization approach with the Virtual Image Space Component Iterative Technique, which combines image

DOI: 10.1201/9781003667926-6

regularization with a linearized forward solver to identify unknown components. As compared to static regularizations like positivity, better estimations of the unknown components can be obtained by iteratively applying thresholding to the reconstruction. This approach mitigates the challenges of complex implementation and high computational requirements associated with statistical, probabilistic algorithms. Unlike minimization algorithms that operate on continuous functions to match the measured data with a restriction, the constraints of the projection regularization method do not rely on a continuous definition of the matching degree with the image.

This chapter expands upon the nonlinear inverse scattering model developed for UGWT in Chapter 4. The FWI method is merged with an adjustable thresholding regularization method to address challenges associated with limited view UGWT. For the purpose of defect identification, the A_0 mode was employed [94]. This chapter is organized as follows. In Section 6.2, a brief overview of the challenges in the limited view problem is provided, followed by the presentation of the theory of FWI with adaptive thresholding regularization. This method is validated by FE simulations and experiments. The setup is introduced in Section 6.3, and the reconstructions of an irregularly shaped defect as well as two defects are shown in Section 6.4. In 6.5, discussions are carried out to explore the impact of the array length on the reconstructed image. Finally, conclusions are summarized in Section 6.6.

6.2 THEORY

6.2.1 ANALYSIS OF THE LIMITED VIEW PROBLEM

For simplicity, the limited view problem is analyzed within the framework of the Born model under the far-field assumption. Within this framework, the scattered field U_s can be described using the following equation [22,89]

$$U_s \approx -\int_{\Omega} e^{ik(\hat{r}_0-\hat{r})x} O(x)dx, \tag{6.1}$$

where Ω represents the domain of the scatterer and k is the wavenumber. $\hat{r}_0$ and $\hat{r}$ are the directions of the incident illumination and the reception, respectively. $O(x)$ represents the object function, which mathematically depicts the scatterer. In the spatial frequency domain, the scattered field can be effectively represented by a Fourier transformation of this object function, denoted as $\tilde{O}$:

$$U_s(\hat{r}_0,\hat{r}) \approx -\tilde{O}[k(\hat{r}_0-\hat{r})]. \tag{6.2}$$

This suggests that the signal obtained from each source-receiver pair corresponds to the Fourier component of the object function at $k(\hat{r}_0-\hat{r})$. As illustrated in Figures 6.1(a) and (b), the ideal reconstruction of the object can be achieved by using information from all possible illumination and reception angles for all frequencies within a disk [22]. The impact of the limited view is that it leaves regions where the estimation of the Fourier transform of the object cannot be obtained, effectively introducing a low-pass filter for the areas lacking data. The typical limited view problems with two parallel linear arrays measuring the transmission data and their filtered

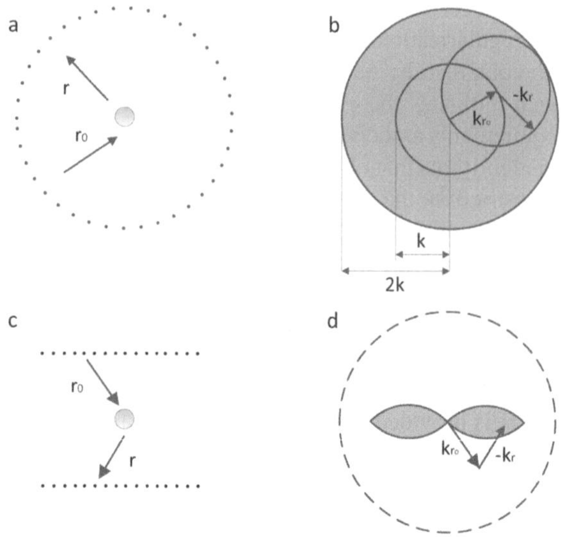

Figure 6.1 2D Fourier transform based on Born approximation. (a) The full view array and (b) the corresponding Fourier components obtained from each layout; (c) a pair of parallel linear arrays with a source array above the defect and a receiver array below it; and (d) the corresponding Fourier components [171].

wavenumber spectra are shown in Figures 6.1(c) and (d). More detailed discussions can be found in [55, 89].

Therefore, the central challenge in limited view reconstruction is to accurately estimate the missing components, which could be achieved using prior information. For instance, the reconstructed values of the scatterer should be equal to or larger than the background value [230].

6.2.2 REGULARIZATION TO SYNTHESIZE UNKNOWN COMPONENTS

To minimize artifacts resulting from limited view scenarios, the regularization approach based on prior information via the iterative algorithm is employed to obtain more appropriate data for unknown components. An illustration of this process can be found in references [89, 230], as depicted in Figure 6.2. The regularization assumption is made that scatterer values are the same or greater than the background values.

Based on the far-field assumption, the known Fourier components can be directly obtained from the data, as shown in Figure 6.2(a), which is similar to Figure 6.1(d). The values outside the known components are set to zero. The image shown in Figure 6.2(b) can be obtained directly from the Fourier components shown in Figure 6.2(a) by taking inverse Fourier transform. Since the unknown components are assigned a value of zero, the resulting reconstruction becomes a low-pass filtered version of the true image, leading to the appearance of “x”-shaped artifacts [55].

Subsequently, the reconstruction is regularized by setting negative values to the background value of zero. This regularization is followed by taking the forward Fourier transform of the resulting image, as illustrated in Figures 6.2(c) and (d), respectively. It can be seen that the previously unknown components, which were assigned zero values in the wavenumber space image, now become non-zero values, indicating the acquisition of additional information from the missing components. Next, the estimations of the unknown components (see Figure 6.2(e)) are combined with the known components. The process is iterated until convergence is achieved.

6.2.3 ADAPTIVE THRESHOLDING REGULARIZATION METHOD

In many cases, although artifacts associated with the limited-view problem may obscure lower amplitude features, peak features are often relatively well reconstructed [55], as demonstrated in Figure 6.3(a). This characteristic can be effectively exploited by an adaptive thresholding approach [99]. In this method, a threshold is applied to the image, which is set below the level of the artifact but above the level of the peak feature, as shown in Figure 6.3(b). Below this threshold, all values are preserved, while values above the threshold are assigned to the background. This process enables the population of unmeasurable components due to limited view angles. The additional information from the missing angles, obtained through this regularization technique, is combined with the measured data, leading to a reduction in artifacts above the threshold, as shown in Figure 6.3(c). As artifacts diminish in subsequent iterations, the threshold level can be gradually increased towards the background level until most of the artifacts are eliminated.

Note that the threshold method can be applied in both cases where the reconstructed value is either above or below the background. The threshold can be defined as follows [99]

$$\mathrm{m}_{reg}(x) = \begin{cases} \mathrm{m}(x) & if \frac{|\mathrm{m}(x)-\mathrm{m}_0|}{\max|\mathrm{m}(x)-\mathrm{m}_0|} < l_e \\ \mathrm{m}_0 & \text{otherwise}, \end{cases} \tag{6.3}$$

where $\mathrm{m}(x)$ is the model parameter at x, m_0 being the level of the background, and l_e is the threshold fraction.

6.2.4 IMPLEMENTATION OF THE REGULARIZATION METHOD IN GUIDED WAVE TOMOGRAPHY

Figure 6.4 outlines the flowchart of the regularization method in combination with the FWI algorithm. Initially, the unmeasurable components are set to zero. The image is obtained by using the measured, limited view data based on the FWI algorithm. Then, regularization based on the adaptive thresholding method is applied. At each threshold level, a non-linear forward solver of the current regularized image is used to generate unknown data in the FWI algorithm, and then the inversion is performed

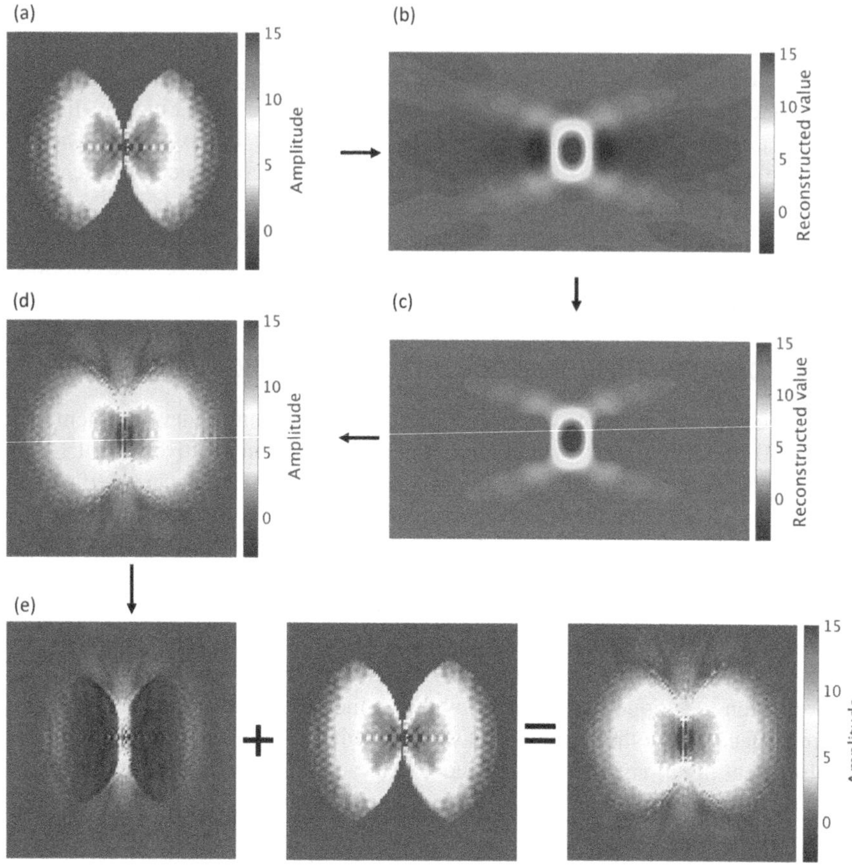

Figure 6.2 Applying the approach used by Huthwaite [89] and Sung et al. [230] to synthesize unknown Fourier components by regularization. (a) Fourier components from a limited view, (b) the reconstruction obtained by using the Fourier transform, followed by the application of regularization in (c), and (d) additional information is obtained through regularization. The estimations of the unknown components in (e) are then combined with the known components [171].

to reconstruct from that. The information included in the unknown components contributes to the improvement in the image caused by the regularization. Thus, the solution incorporates this additional information into the measured, limited-view data, which can achieve a better estimation of the image for the next iteration. This process is repeated until convergence is achieved. The mean square error (MSE) can be used

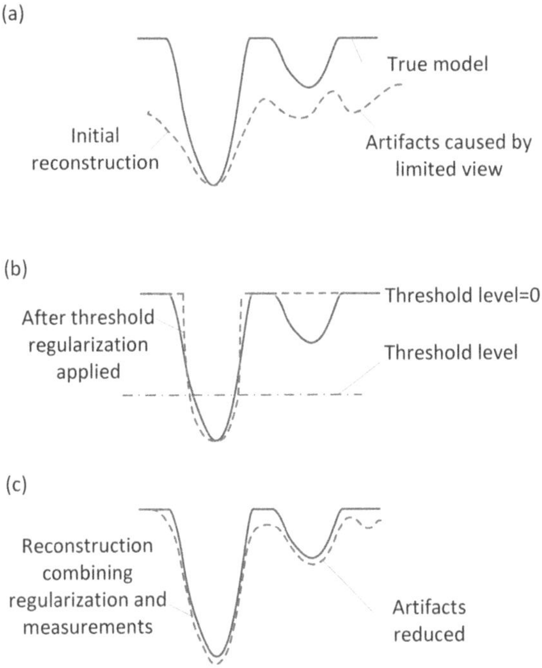

Figure 6.3 1D example illustrating the schematic outline of the method of thresholding regularization. (a) The standard limited view reconstruction, (b) a threshold used in the image, and (c) the next regularized reconstruction [171].

as a criterion of convergence, similar to references [41,99]

$$MSE_{(FWI)} = \sqrt{\frac{\sum_{pi=1}^{N_{pi}} (T_{pi}^{(i)} - T_{pi})^2}{\sum_{pi=1}^{N_{pi}} T_{pi}^2}} \tag{6.4}$$

where T_{pi} is the original thickness at pixel pi, $T_{pi}^{(i)}$ represents the reconstruction of the thickness at pixel pi for the ith iteration, and N_{pi} represents the total number of pixels in the reconstructed image. The value of MSE decreases as the number of iterations increases and eventually converges to a single value. Experience has shown that convergence for a single frequency can be achieved within 10 to 20 iterations.

It is crucial to establish an appropriate threshold fraction for regularization. In this chapter, the threshold fraction can be expressed as $f^{(n+1)} = p(f^{(n)} + q) - q$, similar to the literature [99]. The parameter p determines the degree of threshold increase at each iteration, and q represents the asymptotic limit. For this chapter, values of p = 0.8, q = 0.2, and $f^{(0)}$ = –0.6 were utilized. The background level is set to zero, meaning that the threshold fraction below this level is negative. The initial value $f^{(0)}$ can be adjusted according to the specific case. Initially, the standard

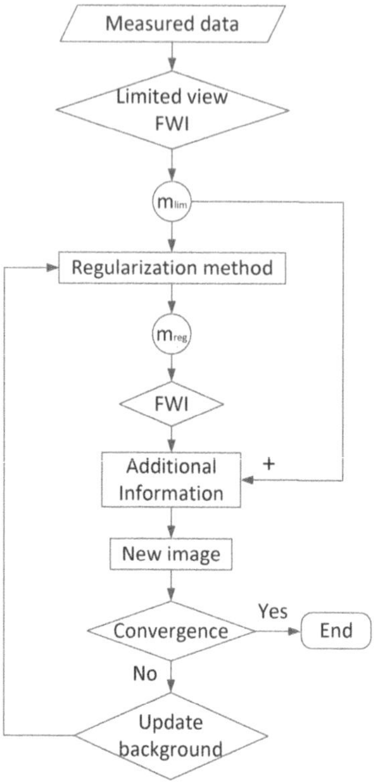

Figure 6.4 The flowchart of regularization with the FWI algorithm for a limited view configuration [171].

limited view FWI algorithm was implemented using multiple frequencies. Based on the results obtained from the final frequency, we chose to perform each regularization iteration for several FWI iterations. In this chapter, after one regularization iteration, approximately 10 FWI iterations are executed to update the background after each regularization iteration. This iterative process continues by repeatedly applying the regularization until convergence is achieved.

6.3 VALIDATION METHODS

6.3.1 DEFECTS

In this chapter, two types of defects, irregularly shaped defects and axisymmetric flat-bottom circular ones, are considered on an aluminum plate (see Table 4.1) with dimensions of 1100 mm × 1100 mm × 10 mm. The first type is irregularly shaped defects, with the largest size being approximately 100 mm and a depth of 3 mm. The second type is axisymmetric flat-bottom circular defects, which can be characterized by their surface diameter, bottom diameter, and defect depth [18]. In the case of two

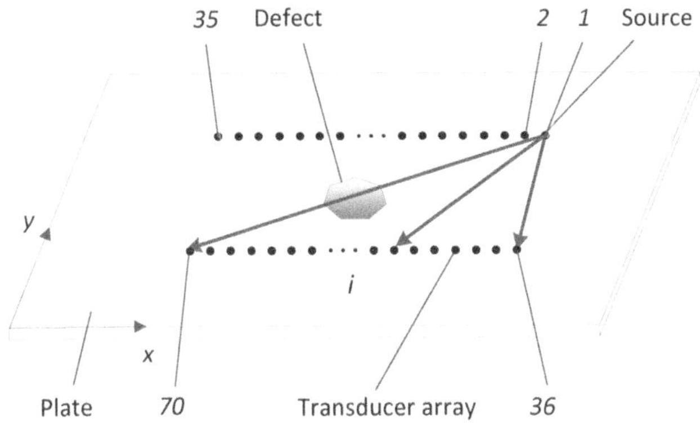

Figure 6.5 The configuration of a pair of parallel linear transducer arrays for UGWT on a plate [171].

flat-bottom defects, each defect is offset by 40 mm from the center of the plate in the horizontal direction. The larger defect has a surface diameter of 60 mm, a bottom diameter of 44 mm, and a depth equal to 50% of the nominal plate thickness. The smaller defect has surface and bottom diameters of 42 mm and 34 mm, respectively, and its thickness reduction is 20%.

6.3.2 NUMERICAL MODELING

To validate the proposed algorithm, FE simulations were conducted for the two aforementioned cases. The plate was modeled using cubic-shaped brick elements with an element size of 1 mm. To minimize reflections from the plate edges, absorbing layers were applied [168]. The simulation included a pair of parallel linear transducer arrays, as shown in Figure 6.5. A 5-cycle Hanning-windowed tone-burst signal with a frequency of 50 kHz was used as the input signal. For a specific excitation point on one array, the A_0 mode was generated by applying an out-of-plane force, and the resulting wavefields were measured by all receivers on the other array. The configuration of the transducer arrays slightly differs between the two cases. For the irregularly shaped defects, each array consists of 35 different generator/monitor positions, equally spaced by 20 mm ($\simeq 0.6\lambda$ at 60 kHz), with a spacing of 150 mm between the two parallel linear transducer arrays. In the cases involving two defects, each array comprises 17 generator/monitor positions, spaced by 40 mm ($\simeq 1.2\lambda$ at 60 kHz), and the space between the two arrays is 200 mm.

6.3.3 EXPERIMENTAL PROCEDURES

The experimental setup is presented in Figure 6.7. The defects were introduced into the plate using a CNC milling machine, maintaining the same dimensions as in the FE simulations. The A_0 mode was excited using a PZT transducer (Panametrics

V1011) at one position. The transducer was coupled to the plate through a small area using an aluminum disk with a diameter of 10 mm and a thickness of 0.5 mm [186]. The displacement normal to the surface of the plate was measured using a Polytec OFV-505 laser vibrometer. During the measurement, a 5-cycle Hanning-windowed tone-burst signal centered at 50 kHz was generated using a Tiepie Handyscope HS3.

To prevent reflections from the plate edges, only transmitted signals were measured. As a result, the measurements were performed along a single linear transducer array, as shown in Figure 6.7. By repeating this process, a matrix of size 70×35 was obtained for the signals corresponding to the irregularly shaped defect, while a matrix of size 34×17 was generated for all source-receiver combinations related to the two flat-bottom defects. An SNR greater than 35 dB was achieved by averaging the signals over 100 repetitions. Furthermore, unwanted parts of the measured signals were eliminated using a gating function [97]. Prior to using these signals in the inversion process, calibration of the obtained signals was performed using a similar approach as outlined in Section 4.3.3.

6.4 RESULTS AND ANALYSIS

6.4.1 IRREGULARLY SHAPED DEFECT

Figure 6.6(a) presents the original model of an irregularly shaped defect used in the experiment. A pair of parallel linear transducer arrays, each containing 35 transducers separated by 20 mm, was employed. The defect has a complex shape with smooth variations in thickness. The largest size of this defect is about 100 mm, and it has a depth of 3 mm. Three sequential frequencies of 35, 46, and 60 kHz were used in the standard limited view FWI algorithm, with 40 iterations per frequency. Following the results obtained from the 60 kHz, regularization was applied. To update the background, 10 FWI iterations were performed. As the artifacts diminished with an increasing number of iterations, the threshold level can be slightly increased. This process is repeated until convergence was achieved. The computations were conducted on an HP Z820 workstation equipped with a 1-core processor and 128 GB of memory. The entire process took approximately 2 hours to generate the regularized results presented in Figure 6.6(c).

Figures 6.6(b)-(e) share the same color bar and scales as Figure 6.6(a), allowing for direct comparisons. It can be seen from Figure 6.6(b) that the standard FWI limited view reconstruction based on the FE simulation shows a slight elongation in the shape of the defect, and clear "x"- shaped artifacts are also visible. Compared to the full view configuration, the relatively small range of viewing angles can slightly distort the defect. However, the overall quality of the resolvable area is generally not degraded. The diagonal artifacts observed are a common result in limited view configurations [134]. These artifacts arise because the unknown components are set to zero in the standard limited view reconstruction, resulting in a filtered version of the true image and the introduction of artifacts. Some artifacts observed near the transducer arrays are primarily caused by undersampling of the wavefield, as the spacing of the transducer locations falls below the Nyquist requirement of half

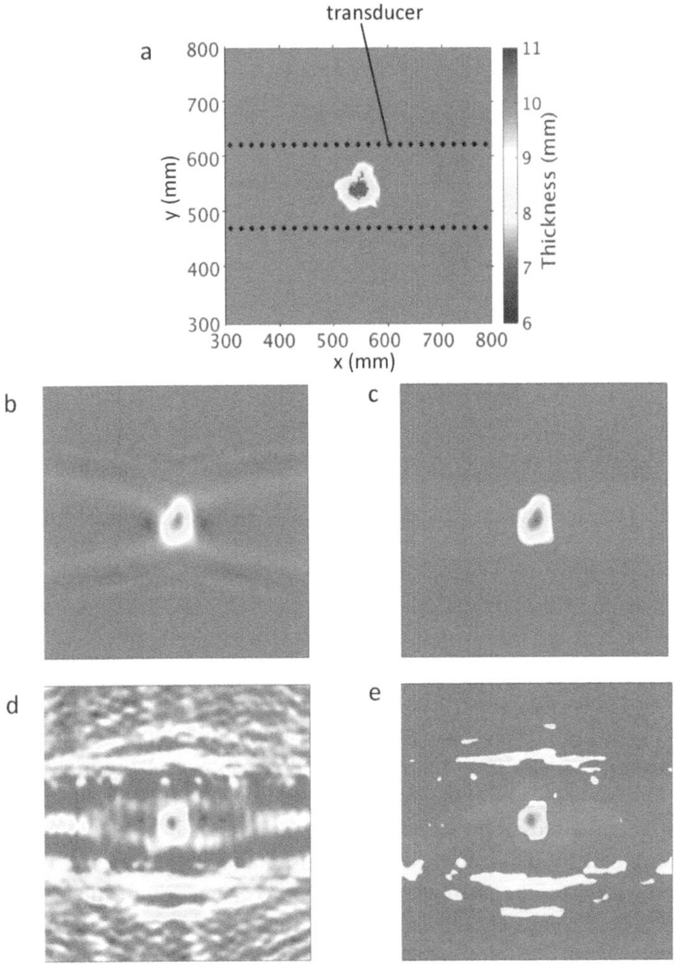

Figure 6.6 Reconstructions of an irregularly shaped defect using a pair of parallel linear transducer arrays. (a) Original model, (b) the standard FWI limited view reconstruction, (c) the regularized reconstruction based on the FE model, (d) the standard limited view reconstruction, and (e) the regularized reconstruction in the experiment [171].

wavelength [170]. Figure 6.6(c) shows the image after regularization which slightly improves the defect representation, and more importantly, artifacts can hardly be observed. In the experimental results, significant artifacts are observed in the standard limited view reconstruction, as shown in Figure 6.6(d). Along with the undersampling of the wavefield, noise and various uncertainties in the experiment can corrupt the measured data [170]. Despite these challenges, a reasonable estimation of

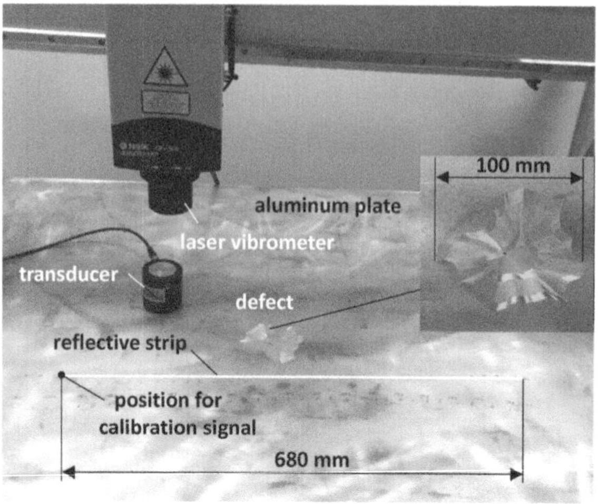

Figure 6.7 The experimental setup for investigating an irregularly shaped defect [171].

the overall shape of this defect can still be achieved. The artifacts are significantly reduced by the regularization method, as shown in Figure 6.6(e).

Figures 6.8(a) and (b) show the cross-sections of the defect at the deepest depths along the horizontal direction in Figure 6.6. The reconstructions using the data from the FE model and experiment slightly underestimate the depths of the defect by approximately 0.1 mm and 0.3 mm, respectively, as shown in Figures 6.8(a) and (b). It is clear that by using the regularization method a slightly better estimation of the defect profile and a reduction in artifacts can be achieved. In Figure 6.8(b), the profiles of the defect by using the experimental data show slightly narrower than the true values, which can be attributed to the limited accuracy of the standard limited view reconstruction and the chosen thresholding level. It is worth mentioning that the reconstructions of the irregularly shaped defect exhibit a relatively low level of artifacts. This is likely due to the use of a relatively large range of viewing angles and small sampling spacing in the measurement, which help mitigate the impact of artifacts [103].

6.4.2 TWO DEFECTS

To demonstrate the suitability of this algorithm for multiple defects, two flat-bottom defects with a relatively large difference in depth are considered in the second model. The modeling is also performed using a pair of parallel linear transducer arrays. The array consists of 17 transducers with a separation distance of 40 mm, as shown in Figure 6.9(a). The larger defect has a surface diameter of 60 mm ($\simeq 1.8\lambda$ at 60 kHz) and a bottom diameter of 44 mm ($\simeq 1.32\lambda$ at 60 kHz), with a maximum depth of 5 mm. The shallower defect has a surface diameter and the bottom diameter of 42 mm ($\simeq 1.3\lambda$ at 60 kHz) and 34 mm ($\simeq 1\lambda$ at 60 kHz) respectively. Its depth corresponds

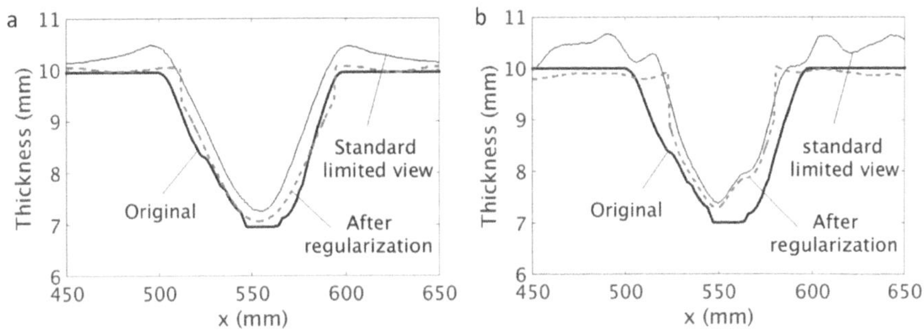

Figure 6.8 Cross-sections of standard limited view FWI and regularized reconstructions of an irregularly shaped defect along the horizontal direction in (a) the FE model and (b) the experiment [171].

to 20% of the thickness of the plate. The choice of frequencies and iterations in this model is the same as in the previous example. For regularization, the initial threshold level is set slightly below the peak value of the shallower defect. This setup allows us to verify the regularization's capability to distinguish between the shallower defect and the artifacts. Consequently, the starting model after the first regularization iteration contains only information about the larger defect and does not include any information about the shallower defect.

The same color bar and scales are used in Figures 6.9(b)-(e). Figures 6.9(b) and (d) show the standard limited view reconstruction based on the FE simulation and the experiment. It can be seen that although the larger defects can be reasonably reconstructed, the smaller defects are almost obscured by artifacts. Significant improvement can be achieved by applying several iterations of regularization, as demonstrated in Figures 6.9(c) and (e). Both defects are clearly reconstructed, with the peak values captured initially and the low contrast features obtained by gradually increasing the threshold level. However, it should be noted that the diameter of the shallower defects appears slightly smaller than the original defect size. This discrepancy is due to the reasons explained previously. There are obvious artifacts in the reconstruction obtained from the experiment, and the main reasons were the undersampling of the wavefield and the presence of noise.

Figures 6.10(a) and (b) present cross-sectional reconstructions across the center of the defects along the horizontal direction. In Figures 6.10(a) and (b), the reconstructions of the larger defects can be accurately observed in both the standard limited view and regularized reconstructions from the FE simulation and the experiment, respectively. The regularized reconstructions of the shallower ones significantly improve the depth estimations compared to the standard limited view reconstructions, despite the sizes being smaller than the original ones. Very few artifacts are visible in the regularized reconstructions, which closely resemble the true values.

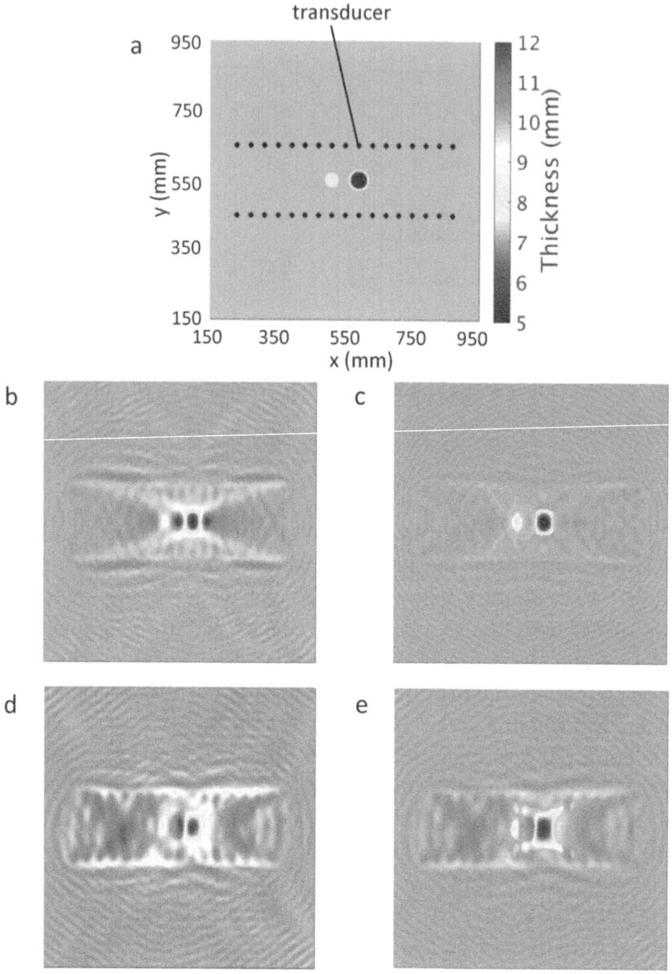

Figure 6.9 Reconstructions of two flat-bottom defects using a pair of parallel linear transducer arrays. (a) Original model, (b) the standard limited view reconstruction, (c) the regularized reconstruction based on the FE model, (d) the standard limited view reconstruction, and (e) the regularized reconstruction based on the experiment [171].

6.5 DISCUSSION

In Section 6.4, the focus of the work was on evaluating the performance of the regularization method for different types of defects with a relatively wide range of viewing angles. However, in practical scenarios, the placement of transducers is often limited to a specific area, resulting in a constrained range of viewing angles. In this section, a single flat-bottom defect was used to evaluate the influence of different array

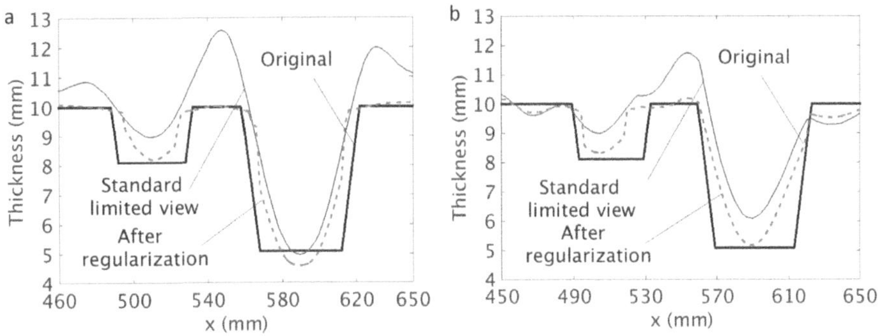

Figure 6.10 Cross-sections of standard limited view FWI and regularized reconstructions of two flat-bottom defects along the horizontal direction in (a) the FE model and (b) the experiment [171].

lengths using data obtained from FE simulations and to demonstrate the effectiveness of regularization. The surface and bottom diameters of this defect are 60 mm and 44 mm, respectively, and its thickness reduction is 50%, as shown in Figure 6.11(a). The varying ratio of the array length L to the fixed distance $D_f = 200$ mm between two arrays and a fixed sampling interval of 40 mm are considered. Figures 6.11(b)-(k) utilize the same color bar and scales as Figure 6.11(a). Figures 6.11(b), (d), (f), (h) and (j) show the standard limited view reconstructions obtained using a pair of linear arrays on each side of the defect. Figures 6.11(c), (e), (g), (i), and (k) demonstrate the reconstructions using FWI with the adaptive thresholding regularization method.

It can be observed from Figure 6.11(b) that when the ratio L/D_f is 4, a good reconstruction of the defect can be obtained using the standard limit view method, although noticeable artifacts are present. A similar result can be achieved when the ratio L/D_f is equal to 3.2, as demonstrated in Figure 6.11(d). However, as the array length decreases, the artifacts around the defect gradually expand. More pronounced distortions in the shape of the defect and underestimations of the peak contrasts can be observed in Figures 6.11(f), (h), and (j). To obtain satisfactory results for this particular defect, the ratio L/D_f needs to be larger than 3.2. Comparing the standard limited view results with the reconstructions after regularization, it is evident that the regularized images achieve significant improvements. They exhibit fewer artifacts and provide better estimations of the contrast of the defects. These improvements are still noticeable even when the ratio is very small, as shown in Figure 6.11(k).

The maximum depth error is a crucial metric in the reconstruction of thickness maps, representing the difference between the reconstructed depth and the true depth divided by the nominal plate thickness [170]. Figure 6.12 illustrates the maximum depth errors for FWI with and without regularization, considering different ratios of L/D_f. It is clear that the maximum depth errors are slightly reduced when regularization is applied, indicating the beneficial impact of the additional information obtained through the regularization method. Moreover, as the length of the array increases, the

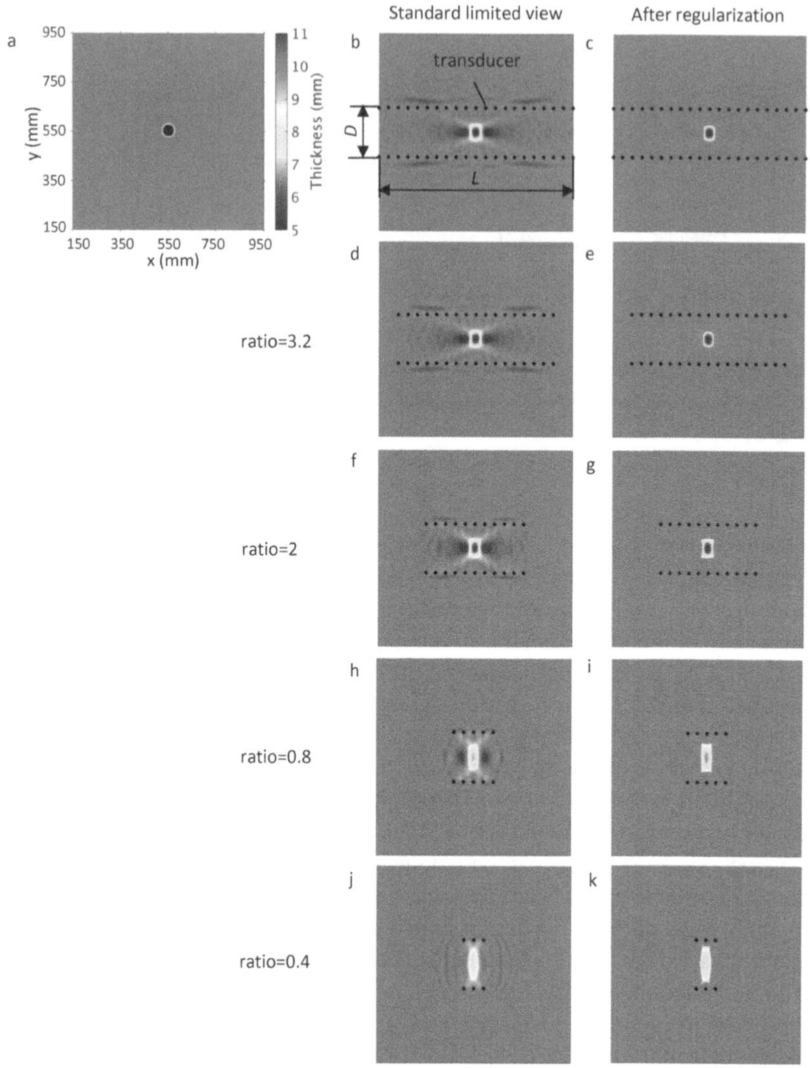

Figure 6.11 Limited view reconstructions of a flat-bottom defect based on the FE model at 60 kHz. (a) The first column shows the original model with a defect depth of 5 mm and a surface diameter of 60 mm; the second column, i.e., (b), (d), (f), (h), and (j), presents the standard limited view reconstructions using FWI; and the third column, i.e., (c), (e), (g), (i), and (k) exhibits the reconstructions obtained using FWI with regularization. In (b) and (c), the ratio L/D_f is 4; (d) and (e) reduce the ratio to 3.2, and (f) and (g) show a reduced ratio of 2. In (h) and (i), the ratio is equal to 0.8, and (j) and (k) reduce that to 0.4 [171].

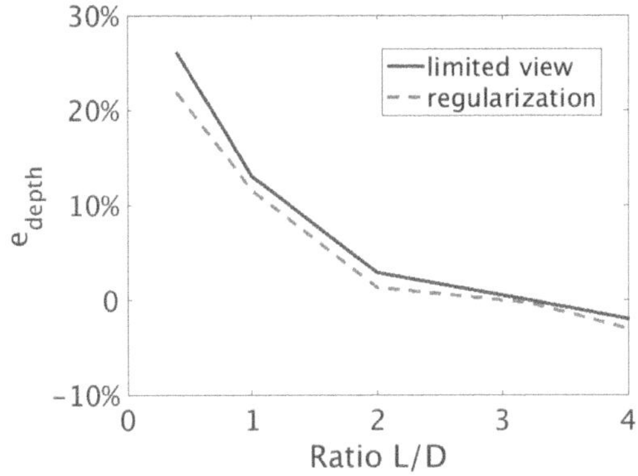

Figure 6.12 The relationship between the ratio L/D_f and the maximum depth error e_{depth} with a single flat-bottom defect [171].

Table 6.1
Comparison of the maximum depth errors (e_{depth}) between the reconstruction with and without regularization [171]

ratio L/D_f	0.4	0.8	2	3.2	4
standard limited view	26%	17.5%	2.9%	0%	-2%
limited view with regularization	21.8%	15.1%	1.3%	-0.3%	-3%

viewing angles become larger, leading to a reduction in reconstruction errors. The difference in the maximum depth errors between FWI with and without regularization, as extracted from Figure 6.12, is summarized in Table 6.1. The table highlights the improvement achieved through the regularization approach, confirming the trend observed in the cases of irregularly shaped defects and two flat-bottom defects when applying regularization. This part emphasizes the effectiveness of the regularization method in improving the accuracy of the reconstructed thickness maps.

6.6 SUMMARY

This chapter introduced the regularization technique for addressing the challenge of missing components in the FWI algorithm when using a limited view configuration. The regularization method utilized an adaptive thresholding approach to improve the limited view reconstruction. The effectiveness of the FWI algorithm with regularization was demonstrated through simulations and experiments involving irregular defects and two defects. The results clearly showed that the image quality was significantly enhanced by incorporating the additional information obtained through the regularization process with measured data from the limited view configuration. The regularization approach was further evaluated by varying the array length in limited view elastic data, and similar improvements were observed. The limited view configuration, particularly with two circular arrays on either side of the defect region, proved to be highly useful for assessing pipes. This aspect will be further investigated in Chapter 7.

6.7 EXERCISES

1. Please give some examples of typical limited view problems.
2. What is the most prominent problem in reconstruction from incomplete data?
3. How to improve the reconstruction from a limited view?
4. How to utilize the prior information in limited view reconstruction problems?
5. Please analyze the problem of limited view from a mathematical point of view.
6. What is the regularization approach?
7. Please explain the core idea of the adaptive thresholding regularization method mentioned in this chapter.
8. How to implement the regularization method in ultrasonic guided wave tomography?
9. Taking two parallel linear arrays as an example, please analyze the effects of the varying ratio of the array length L to the fixed distance D_f mm between two arrays on the reconstruction.
10. What is the difference between the results of using regularization and not using regularization? Please give the reasons.
11. How will the content of this chapter be applied to pipeline inspection? How does the setup of two circular arrays on either side of the defect region in pipeline inspection relate to the two parallel linear arrays mentioned in this chapter?

7 Applications of Corrosion Mapping

Based on the previous theoretical foundations, this chapter focuses on the application of corrosion mapping. Section 7.1 first considers the leakage of guided wave energy into the liquid, which serves as valuable preparation for the subsequent section on the liquid-loaded plate and pipeline corrosion monitoring system.

7.1 LIQUID-LOADED TANK FLOOR

7.1.1 INTRODUCTION

Large-scale storage tanks are commonly utilized in the petrochemical industry, and the floors of liquid-loaded tanks are prone to corrosion in humid or chemically aggressive environments, resulting in the loss of wall thickness. This chapter extends the application of UGWT to liquid-loaded plates. Compared to the application on free plates, thickness reconstruction of liquid-loaded plates poses additional challenges due to the leakage of guided wave energy into the liquid. However, most existing work on UGWT has focused on waveguides with free boundaries, and addressing waveguides with liquid loading remains challenging due to the strong attenuation of guided waves, resulting in a low SNR. Ultrasonic guided wave attenuation tomography, which considers signal losses due to liquid loading, was used for the localization of defects in the floor of a storage tank [133]. However, obtaining quantitative information on thickness losses has not been successfully achieved.

This chapter introduces the UGWT method with consideration for attenuation to quantify the remaining thicknesses of liquid-loaded plates. The thickness mapping is achieved using FWI, which takes into account higher-order scattering. The attenuation of waves is incorporated through complex velocity in the forward modeling, utilizing a 2D acoustic model approximation. Section 7.1.2 discusses the suitability of low-frequency Lamb modes for UGWT in the presence of liquid loading, while Section 7.1.3 presents the imaging algorithm that accounts for attenuation. The chapter continues with Section 7.1.4, which outlines the FE modeling and experimental procedures conducted. The influence of attenuation on the accuracy of reconstruction, considering various defect widths and depths at different frequencies, is investigated in Section 7.1.5. To validate the effectiveness of the proposed approach, an experiment is conducted in Section 7.1.6, focusing on the reconstruction of an irregularly shaped defect on a water tank floor.

DOI: 10.1201/9781003667926-7

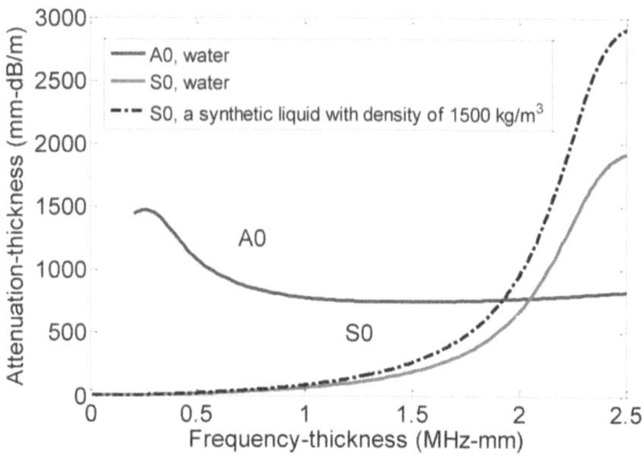

Figure 7.1 The attenuation dispersion curves for an aluminum plate with fluid loading. The dashed dot line represents the attenuation for a synthetic liquid with a density of 1500 kg/m^3 [174].

7.1.2 PROPERTIES OF LOW FREQUENCY LAMB MODES OF LIQUID-LOADED PLATES

The choice of suitable modes for the tomography of fluid-loaded structures relies on the sensitivity of the mode to thickness variations and attenuation caused by fluid loading. The behavior of the fundamental Lamb modes A_0 and S_0 was discussed in Section 1.2.2.1. It was demonstrated that the A_0 mode exhibits better sensitivity to thickness variations compared to the S_0 mode. However, the propagation range of the A_0 mode is significantly limited due to higher attenuation compared to the S_0 mode.

Figure 7.1 illustrates the attenuation of A_0 and S_0 modes when fluid is present on one side of an aluminum plate. It can be clearly seen that the S_0 exhibits significantly lower attenuation in the liquid-loaded plate compared to the A_0 mode below 1.5 MHz-mm. As a result, the S_0 mode is chosen for use in this chapter. To investigate the impact of attenuation on reconstruction accuracy, calculations are also performed for a synthetic liquid with a density of 1500 kg/m^3, which leads to increased attenuation. Further discussion on the effect of attenuation on reconstruction accuracy will be presented in Section 7.1.5.

7.1.3 FWI WITH ATTENUATION

The velocity mapping is generated before being converted into the thickness mapping. This is accomplished using the FWI method, which was extensively discussed in Chapter 4. In the presence of liquid loading, the attenuation of the waves can be presented by complex velocity $\tilde{v}(\omega)$ [239] in the frequency domain FWI, as follows

$$\tilde{v}(\omega) = v(\omega)(1 - \frac{i}{2Q_{atten}}), \tag{7.1}$$

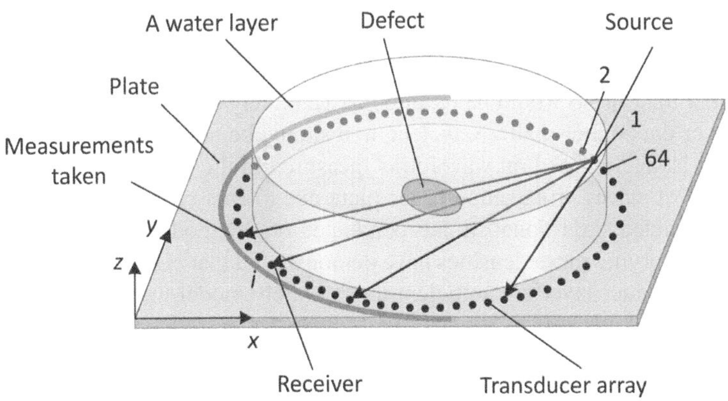

Figure 7.2 Configuration of the fluid-tank model with a defect located at the center of the plate and a transducer array [174].

where v is the real-valued phase velocity, as shown in Figure 1.2(a). Q_{atten} is the attenuation factor and is expressed as

$$Q_{atten}(\omega) = \frac{\omega}{2\alpha_{atten} v(\omega)}, \tag{7.2}$$

where α_{atten} is the attenuation given in Np/m.

7.1.4 TEST APPROACH

7.1.4.1 Configuration

The general configuration of the model and the experiment conducted in this work is depicted in Figure 7.2. It involves a circular water layer with a diameter of 600 mm positioned on the top of a plate. At the center of the plate, there is a surface defect. A circular array with a diameter of 700 mm is placed on the plate, consisting of 64 transducers equally spaced around the circumference. The transducers are located outside the region with water loading. The S_0 mode is excited on each of the transducers and received by the remaining transducers. However, only the transmitted signals are used for the reconstruction process.

7.1.4.2 FE simulations

Hann-shaped defects with different sizes and depths were created in the center of an 1100 mm × 1100 mm × 10 mm aluminum plate (refer to Table 4.1) coupled with a liquid layer. The liquid layer had a bulk modulus of 2.2 GPa, water density of 1000 kg/m^3, and synthetic liquid density of 1500 kg/m^3. The modeling of the problem was performed using Abaqus, and the configuration is illustrated in Figure 7.2.

Since full-scale modeling of the problem is computationally demanding, the model was divided into parts with coarser meshes to optimize computational efficiency. The entire model was meshed with 8-noded linear cubic-shaped elements with a size of 1 mm, resulting in approximately 30 million elements. The selected element size of 1 mm ensures that more than 30 elements per wavelength are used at 150 kHz (λ = 34.4 mm), which is sufficient for the accurate simulation of elastic waves in the plate [98].

While a finer mesh would be required to accurately model acoustic waves in the liquid layer due to the smaller wavelength, it should be noted that the reconstruction procedure is solely based on waveforms measured in the plate. Therefore, only the attenuation of waves propagating in the plate due to energy leakage needs to be accurately modeled. Additional studies conducted on 2D models of plates with water loading and varying mesh densities have demonstrated that eight elements per wavelength in the water layer are sufficient to accurately model the attenuation of plate waves.

To reduce the height of the water layer, an absorbing region with increasing damping was utilized [58]. This approach helps to prevent reflections from the upper surface of the water, which could interfere with the plate waves and complicate the measurements. The plate-water interface was simulated using a surface-based coupling constraint in the out-of-plane direction. Thickness reductions were modeled by reducing the out-of-plane dimension of the elements [90]. To minimize the model size in the plate, the absorbing boundary was employed around the plate, following a similar approach as that used for the water layer, to prevent boundary reflections.

The signals were excited and measured using a 64 transducer array. The input signal was a 5-cycle Hanning-windowed tone-burst centered at 150 kHz. At this frequency, the S_0 mode exhibits significant in-plane and out-of-plane displacement components. To selectively excite the pure S_0 mode, symmetric out-of-plane forces were applied on the top and bottom surfaces of the plate in the out-of-plane direction, with respect to the mid-plane of the plate [92]. During reception, the S_0 waves were obtained by calculating the difference between the out-of-plane components measured at both surfaces of the plate.

7.1.4.3 Experimental setup and procedures

The experimental setup is presented in Figure 7.3. The experiments were conducted on an aluminum plate with sizes of 1100 mm × 1100 mm × 10 mm. An irregularly shaped defect was machined in the center of the plate. The tank wall on the plate consists of a 2 mm thick aluminum sheet that has been rolled into a cylinder with an inner diameter of 600 mm. It is attached to the plate using silicon, and the tank is filled with water up to a level of 200 mm. A PZT transducer (Panametrics V1011) was used to excite the plate in the out-of-plane direction. A 5-cycle Hanning-windowed tone-burst signal centered at 150 kHz was generated by the RITEC RAM-5000 high-power ultrasonic pulser (Ritec Inc., USA). The out-of-plane components of the waves at the plate surface were captured using a laser vibrometer. A reflective strip was applied to the surface of the plate to reflect the signals back to the laser receiver.

The measurements were repeated for 64 excitation locations on an array, and the signals underwent post-processed before being used in the imaging algorithm. In the experiment, both the S_0 and A_0 modes were generated under out-of-plane excitation.

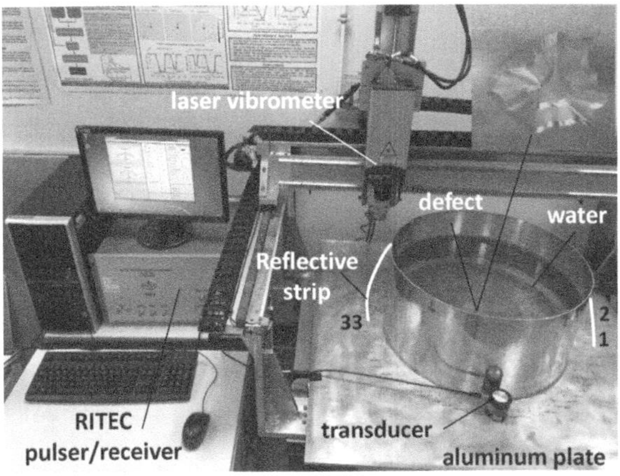

Figure 7.3 The experimental setup. Note that an irregularly shaped defect was machined in the aluminum plate [174].

However, they could be separated by applying appropriate time gating. Figure 7.4(a) displays a contour plot of the typical out-of-plane displacement signal measured over the receivers. The first arrival of the S_0 mode was separated from the signals by removing unwanted components using a gating function, as shown in Figure 7.4(b). The gating function is a rectangular window smoothed with rising and falling sinusoids at the edges on both sides. The length of the gate varied with the propagation distance due to the dispersion of the S_0 mode. The starting point of the gate was determined by the arrival of the S_0 mode, and the gate ended at the arrival of the slower A_0 mode. All signals after time gating were calibrated before being used in the inversion, following a similar method introduced in Section 4.3.3.

7.1.5 EFFECT OF THE ATTENUATION ON THE RECONSTRUCTION ACCURACY

As explained in Section 7.1.2, the S_0 mode in a liquid-loaded plate experiences significant attenuation at frequencies where its dispersive nature can be utilized for thickness reconstruction. To assess the impact of this phenomenon, a series of experiments are conducted with varying defect widths and depths at different frequencies. These results are then compared with the reconstructions obtained from a free plate and a liquid-loaded plate where the attenuation is not taken into account in the FWI algorithm. The objective is to determine the extent of improvement in reconstruction accuracy when attenuation is considered in the inversion process.

Figure 7.5 presents the plate thickness reconstructions of a Hann-shaped defect located at the center of a circular array, based on the FE modeling data. The defect has a surface width of 100 mm and a depth of 3 mm in a 10 mm thick aluminum plate. In the inversion process, sequential frequencies of 130 kHz, 150 kHz, and

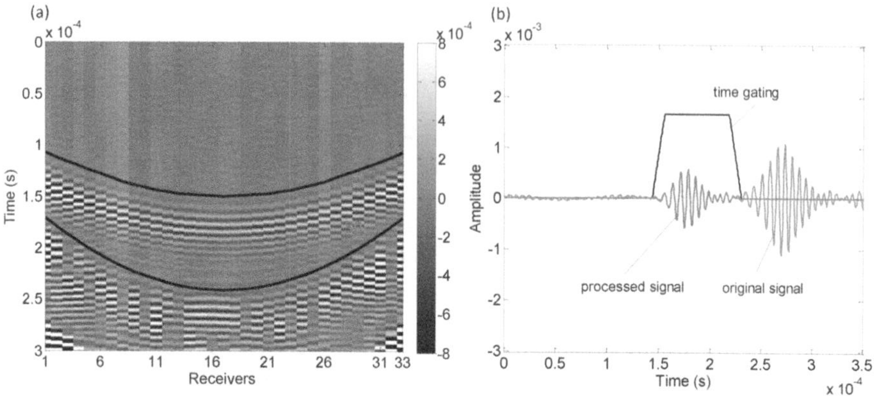

Figure 7.4 (a) Contour plot of measured out-of-plane displacement generated by the transducer shown in Figure 7.3. The solid curves represent the adaptive time gating applied based on different propagation distances. (b) Time trace from one of the receivers, and the signal after time gating [174].

180 kHz are utilized. The homogeneous background is used as the starting model at the lowest frequency, and 20 iterations are implemented at each frequency. It can be seen from the figures that the defect is successfully reconstructed at the center of the plate in all cases, and the contrast is slightly improved at higher frequencies. In the reconstructions of the defects on water-loaded plates, the areas with water loading remain free of artifacts in most images. However, artifacts are present at the location of the array, and their visibility increases with higher frequencies. This can be attributed to the combination of undersampling of the measured wavefield in the circular array and the larger discrepancy between the amplitudes of the modeled and observed wave data [170,218]. Significantly more pronounced artifacts are observed when the inversion is performed without considering the attenuation. It should be noted, however, that even when the attenuation is taken into account in the inversion algorithm, artifacts do not completely disappear. This suggests that the attenuation used in the inversion may differ from the actual attenuation present in the FE data, resulting in some undesired data residuals during the inversion process.

Figure 7.6 presents the cross-sections of the reconstructions from Figure 7.5 along the central line of the plate at different frequencies of 130, 150, and 180 kHz. At 130 kHz and 150 kHz, all the reconstructions show a slight underestimation of the true defect depth. However, at 180 kHz, a slight overestimation of the defect depth can be observed. This discrepancy can be attributed to the fact that the 2D acoustic model used in the inversion is an approximation of the 3D wave propagation model, leading to some loss of accuracy in the inversion [90]. Furthermore, it can be seen that at low frequencies with small attenuation, the differences between the reconstructions obtained from different approaches are minimal. However, as the attenuation increases with frequency, the reconstructions obtained without considering the attenuation

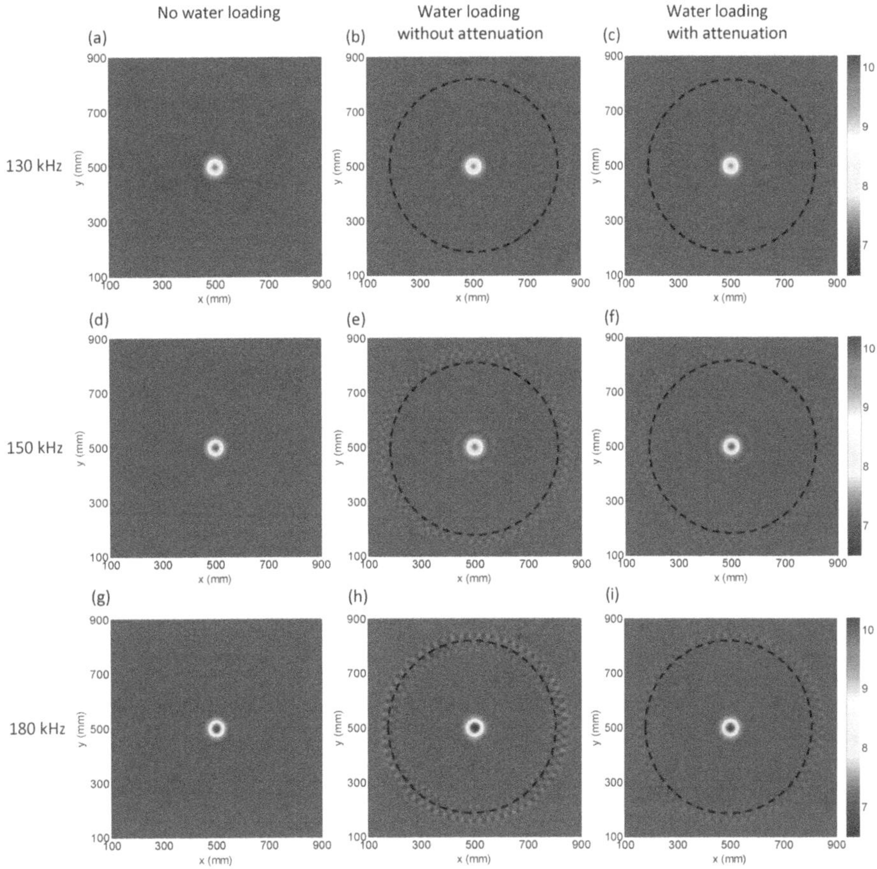

Figure 7.5 Thickness reconstructions of a Hann-shaped defect with a surface width of 100 mm and a depth of 3 mm in a 10 mm thick aluminum plate at different frequencies from the FE model. Each row represents the reconstruction results at a specific frequency, with the first row corresponding to 130 kHz, the second row to 150 kHz, and the third row to 180 kHz. Each column represents the reconstruction results in different scenarios: free plate, water-loaded plate with attenuation considered, and water-loaded plate without considering attenuation. The dashed curves indicate the position of the tank wall [174].

deviate more from the true depth. In contrast, the reconstructions obtained from the inversion with attenuation consideration for both the defect on a free plate and on a water-loaded plate align well with each other.

Figure 7.7 shows the maximum depth error of a Hann-shaped defect with a surface width of 100 mm and a depth of 3 mm as the frequency varies. The results include liquid-loaded plates with varying levels of attenuation to demonstrate the impact of different attenuation levels on reconstruction accuracy. Across all the results,

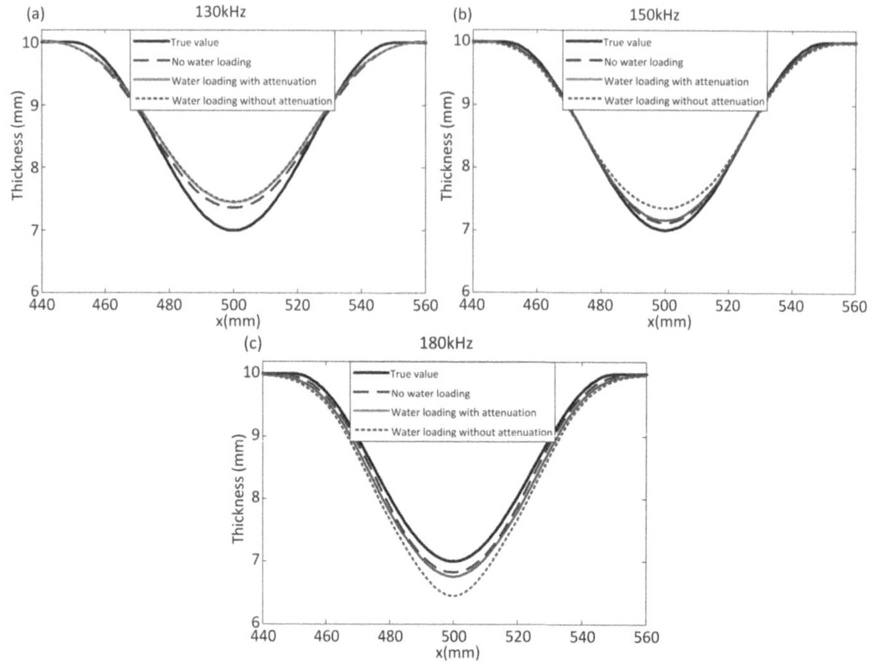

Figure 7.6 Cross sections of the thickness reconstructions from Figure 7.5 along the center line at different frequencies of (a) 130 kHz, (b) 150 kHz, and (c) 180 kHz [174].

similar trends can be observed. At lower frequencies, the maximum defect depths are slightly underestimated (negative value). As the frequency increases, the reconstructed depths show an overestimation (positive value). At lower frequencies where the attenuation is small, the reconstructions with and without considering attenuation in the inversion are very similar. However, as the frequency increases and the attenuation becomes more significant, the differences between the two approaches become more pronounced. For instance, at 150 kHz, the reconstructions that account for attenuation exhibit results that are closer to the true depths. Conversely, at higher frequencies where the reconstructed results show an overestimation of the wall loss, the errors are larger for the cases where the attenuation is not considered. Importantly, it can be clearly seen that the errors increase as the attenuation of the liquid becomes larger. This error is caused by the increase of the attenuation which increases the differences between the amplitudes of the waves that 1) propagate directly from the source to the receiver and 2) are scattered from the defect, compared to the plate without liquid loading. Additionally, this error increases with increasing propagation distances. Overall, when attenuation is incorporated into the reconstructions, the results closely resemble those obtained from the free plates. Only at 160 kHz do the results show some deviations, which may be attributed to the combined effect of

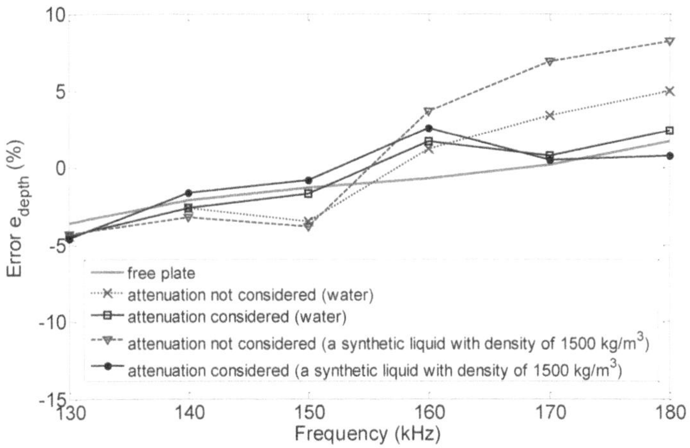

Figure 7.7 The maximum depth error of a defect with a width of 100 mm and a depth of 3 mm at varying frequencies. The reconstructions of a free aluminum plate, water-loaded plate and higher attenuation liquid-loaded plate with and without the consideration of the attenuation in the FWI algorithm [174].

attenuation and the density correction for the S_0 mode. The S_0 mode is more sensitive to errors than the A_0 mode given the nature of the dispersion curves, and applying density correction to the FWI reconstruction with the consideration of attenuation at certain frequencies can affect the reconstruction accuracy [93].

The accuracy of the thickness reconstructions in water-loaded plates, considering attenuation in the inversion, is investigated for different defect widths and frequencies with a fixed depth of 3 mm, as shown in Figure 7.8. The inversion process was conducted over sequential frequencies from 130 kHz to 180 kHz, with a frequency step of 10 kHz. The inversion was started from the homogeneous background model at 130 kHz, with 20 iterations at each frequency. The surface width of the defect was varied from 40 to 120 mm with a step of 20 mm. Several trends can be observed from the results. At a fixed frequency, the depth of defect tends to be underestimated for smaller widths. As the width increases, the reconstruction accuracy improves. Conversely, at a fixed width, the accuracy of the reconstruction improves as the frequency increases. The maximum depth error in the reconstruction of water-loaded plates with the consideration of attenuation aligns well with previous findings on free plates [92, 172]. These findings demonstrated that higher frequencies lead to better resolution due to the smaller wavelengths associated with higher frequencies, leading to better accuracy of the reconstruction.

Figure 7.9 shows the error across a range of frequencies and different depths of the defect in the water-loaded plate with the consideration of attenuation when the surface width is fixed at 100 mm. At lower frequencies, the depth of defect is underestimated, and the error has an increasing trend with the increase of the depth.

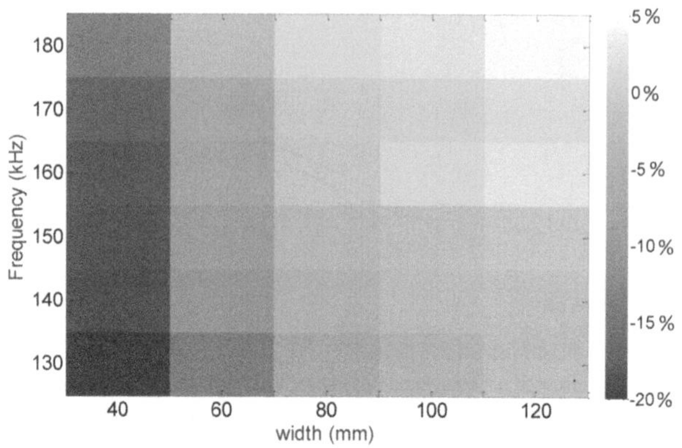

Figure 7.8 The maximum depth error of a Hann-shaped defect with a fixed depth of 3 mm and varying surface widths across different frequencies in water-loaded plates with the consideration of the attenuation with the S_0 mode [174].

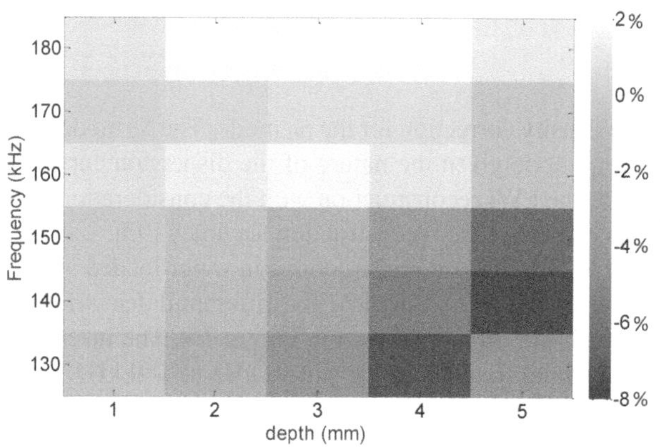

Figure 7.9 The maximum depth error of a Hann-shaped defect with a fixed width of 100 mm and varying depths across different frequencies in the water-loaded plates with the consideration of the attenuation with the S_0 mode [174].

As the frequency increases, the depth error becomes smaller due to the resolution improvement.

In summary, the results have shown that attenuation influences the accuracy of thickness reconstruction. The inclusion of higher attenuation in the inversion process improves the accuracy of thickness reconstruction, particularly when using the S_0 mode at higher frequencies.

7.1.6 EXPERIMENTAL VALIDATION

Figure 7.10(a) shows the original model of an irregularly shaped defect located at the center of a circular array consisting of 64 transducers. This complex-shaped defect has a maximum surface width of approximately 100 mm ($\simeq 2.9\lambda$ at 150 kHz), a flat bottom width of around 25 mm ($\simeq 0.73\lambda$ at 150 kHz), and a defect depth of 3 mm. The reconstruction algorithm, which considers the attenuation of the waves due to liquid loading, was applied. The homogeneous background was used as the starting model, and the inversion was performed at a single frequency of 150 kHz with 20 iterations. It should be noted that more accurate results can be obtained at higher frequencies. However, due to the limitations in propagation distance, it was challenging to separate the S_0 and A_0 modes at higher frequencies, and therefore the reconstruction was conducted at 150 kHz. Figure 7.10(b) presents the reconstruction of the complex defect from the experimental data. It can be seen from the image that the defect is very well reconstructed. The zoomed-in views of defects from Figures 7.10(a) and (b) are depicted in Figures 7.10(c) and (d), respectively, showing good agreement with each other. Notable artifacts can be observed in the image obtained from the experiment, which can be attributed to factors such as undersampling of the wavefield, measurement noise, uneven transducer coupling resulting in unwanted directionality, and additional scattering from the attached tank wall.

Figures 7.11(a) and (b) display the cross-sections through the horizontal and vertical lines, respectively, of the irregularly shaped defect shown in Figures 7.10(c) and (d). It can be seen that the reconstructed profile closely matches the true profile, and the deepest depth of the defect is accurately captured. It is worth noting that a slight overestimation of the maximum defect depth can be observed, which differs from the prediction shown in Figure 7.7 for a Hann-shaped defect of similar size. This discrepancy is primarily attributed to the influence of the bottom profile on the accuracy of the reconstruction. As discussed in [172], when the bottom width of a flat-bottom defect is small, the depth estimation resembles that of a Hann-shaped defect and leads to an underestimation of the defect depth. Conversely, when the bottom width increases, the reconstruction overestimates the defect depth. The reasons have been explained in [93, 172].

These above results, along with the limited view tomography presented in Chapter 6, will provide a strong foundation for the subsequent section on corrosion monitoring of liquid-loaded pipelines.

7.2 PLATE AND PIPELINE CORROSION MONITORING SYSTEM

7.2.1 BACKGROUND

Corrosion is a complex and dynamic process that can be influenced by various factors, including time and process variables. As a result, the monitoring of corrosion processes has become highly sought-after to improve the safety and profitability of petrochemical plants. Traditional off-line inspections, which involve relatively long time intervals, are not capable of detecting unforeseen changes that may have occurred during the progression of corrosion [194]. In contrast, ultrasonic techniques

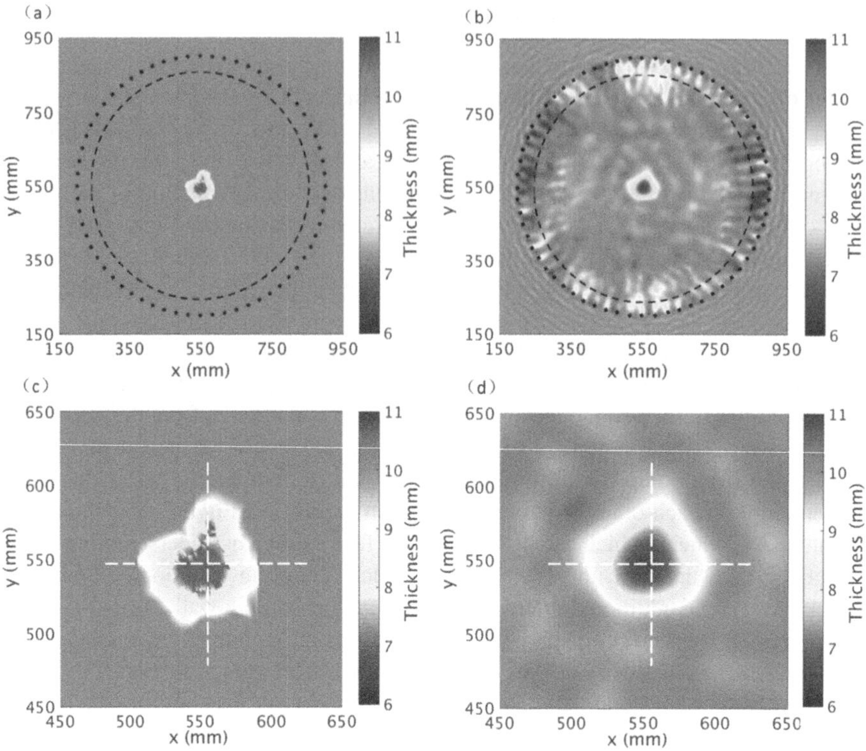

Figure 7.10 FWI thickness reconstructions of an irregularly shaped defect at 150 kHz. (a) Original model, and (b) the experiment; (c) and (d) zoomed-in images of the defect from (a) and (b), respectively. The dotted curve represents the position of the transducer array, and the dashed curve represents the tank wall [174].

can offer a non-intrusive and real-time means for the online monitoring of corrosion [254, 265].

To monitor average thickness loss over a moderate area, a few permanently installed point thickness gauges have been used, providing a high probability of detection [35, 36]. However, when it comes to detecting non-uniform and highly localized defects, a significant increase in the number of transducers is required. In such cases, area monitoring systems (e.g., the field signature approach [227] or dispersive guided wave velocity measurements [73, 101]) become more appealing. However, it's important to note that these area monitoring systems are only sensitive to the average wall thickness loss over the monitored area. Therefore, they may fail to detect or underestimate the severity of non-uniform corrosion if the wall loss is not uniform, as the average thickness loss is taken as representative. In cases of severe wall loss over a large area coverage, the UGWT system has demonstrated the capability and

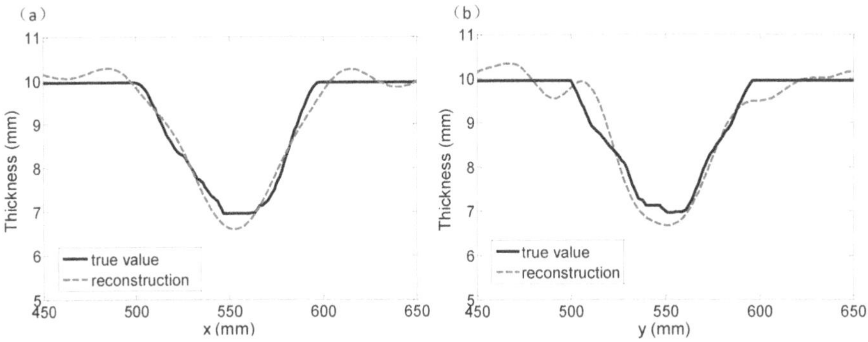

Figure 7.11 Cross sections of the thickness reconstruction of the irregularly shaped defect along the horizontal and vertical line in the experiment from Figure 7.10(c) and (d) at 150 kHz, respectively [174].

sensitivity in mapping non-uniform and localized corrosion damages [18, 46, 97, 275].

An UGWT system typically consists of an array of transducers, a data acquisition system, data processing techniques, and a reconstruction algorithm. A successful UGWT system should possess good sensitivity to corrosion damages, high localization accuracy, high-quality imaging, and a low transducer density [44]. Ultrasonic guided wave-based NDT systems utilizing arrays of electromagnetic acoustic transducers (EMATs) have been used for corrosion monitoring of pipes and plate-like structures [248]. For instance, ultrasonic computerized tomography techniques have been employed to process the received signals from EMAT arrays encircling the pipe, generating thickness loss maps [100, 207]. However, most EMATs have relatively low SNR, and their performance relies on the electromagnetic properties of the tested objects, such as the presence of a highly magnetostrictive oxide layer [193]. In contrast, piezoelectric transducers offer higher SNR, relatively lower cost, and lower energy consumption, making them more suitable for long-term monitoring applications [69, 70, 252, 256]. However, previous studies have shown that the scattering from transducers can affect the accuracy of thickness reconstruction and should be minimized to achieve more accurate reconstructions [18]. In this chapter, an optimized low-frequency A_0 mode transducer is designed for the UGWT system, and the impact of transducer array scattering on the accuracy of thickness reconstruction is investigated.

The UGWT system described in this chapter comprises customized piezoelectric transducer arrays, automated signal generation, acquisition and processing, and the FWI algorithm. This setup enables efficient and labor-saving online monitoring of corrosion progression. The performance of the system is evaluated through lab-based forced corrosion experiments. Section 7.2.2 provides a detailed description of the UGWT system, including the array system, data processing method, and transducer scattering considerations. Section 7.2.3 presents information on the forced corrosion experiments and thickness measurements using laser profilometry. The experimental results and discussions are presented in Section 7.2.4. Finally, the conclusions are summarized in Section 7.3.

7.2.2 MONITORING SYSTEM

7.2.2.1 Array system and experimental setup

Two experimental setups were conducted for the UGWT system in this chapter to monitor corrosion in a steel plate and pipe. The first case involved a circular array consisting of 64 piezoelectric transducers uniformly distributed on a steel plate with a diameter of 700 mm, as shown in Figure 7.12(a). The transducer design aimed to generate the A_0 guided wave in the frequency range of 30 kHz to 90 kHz. The design utilized stacked piezoelectric elements to maximize the efficiency of wave generation and reception while minimizing the transducer's dimensions [159]. Piezoelectric ceramic PZT-5A elements with a thickness of 2 mm and a diameter of 10 mm were used. The waveform generator was controlled by an ultrasonic phased array controller (Lecoeur electronique, Chuelles, France). A 5-cycle Hanning windowed tone-burst signal with a central frequency of 60 kHz was chosen for excitation. A typical time trace and frequency spectrum of the received A_0 mode, excited at point 1 and measured at point 17 in a 10 mm thick steel plate, are shown in Figures 7.12(b) and (c), respectively. The 64 transducers were held in place by a support ring to ensure accurate positioning, as shown in Figure 7.12(a). Couplant was used to establish good contact between the plate and each transducer, and weights were evenly distributed on the mounting bracket of the support ring. Springs were fitted between the support ring and each individual transducer. In this experiment, only transmitted signals from the transmission subset were used to avoid reflections from the plate edges [170]. The current system achieved an SNR greater than 30 dB after five averages.

In the first case, a 1500 mm$\times$1500 mm$\times$10 mm mild steel plate (see 4.1) was chosen for the experiment. At the center of the plate, a corrosion site was created, as shown in Figure 7.12(a). Prior to the corrosion process, an artificial defect with a depth of 1 mm and a diameter of 80 mm was milled at the center of the plate. This approach ensures a smooth transition in depth during subsequent corrosion, rather than a sudden change from the top of the plate to the bottom of the defect. The presence of a relatively smooth thickness variation reduces mode conversions, particularly from the A_0 mode to the S_0 mode. This favors the acoustic assumption made in the thickness reconstruction algorithm [90].

In the second case, the experimental setup for corrosion monitoring of the quiescent water-filled steel pipe is shown in Figure 7.13. This experiment was conducted using a 12-inch pipe (outer diameter of 323.85 mm) with a wall thickness of 10.3 mm and a length of 2 m. An irregularly shaped defect was induced in the middle of the pipe through forced electrochemical corrosion. Two ring arrays of transducers were attached on either side of the defect, with a separation distance of 600 mm.

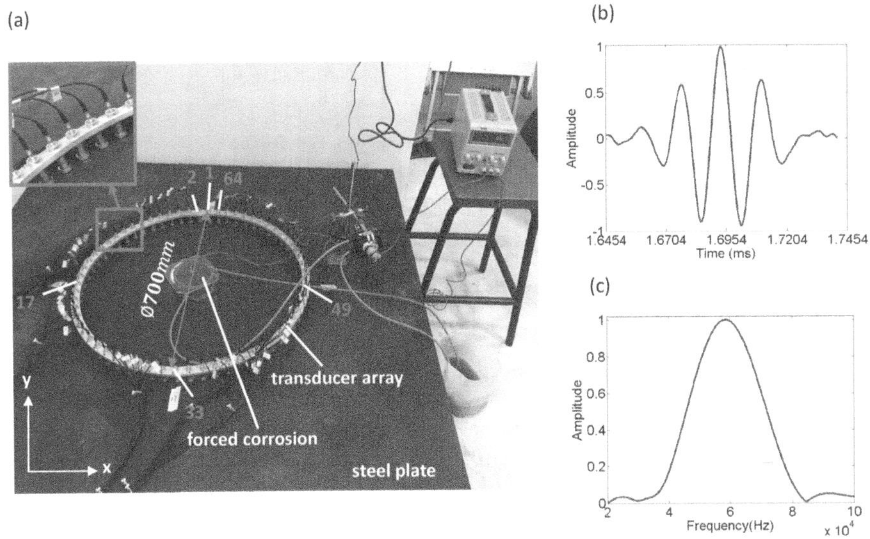

Figure 7.12 (a) Setup for the UGWT measurements on the steel plate. All experiments were conducted at room temperature. (b) Time trace and (c) frequency spectrum of a received signal obtained from the experiment conducted on the steel plate without any defects. The source was at point 1 and the measurement was taken at point 17 [173].

Each array consisted of 24 piezoelectric transducers, which were used to excite the A_0 mode at the center frequency of 60 kHz by applying an out-of-plane force. Two adjacent transducers were positioned approximately 42.4 mm apart in the circumferential direction, which corresponds to 1.25λ ($\lambda = 34$ mm at 60 kHz). This separation distance satisfies the Nyquist requirement of sampling at least half a wavelength, ensuring that the system is not undersampling [218].

In each dataset for the thickness mapping of the pipe, a matrix of 48×24 source-receiver combinations was recorded. It is worth noting that the measurements were initially conducted on a defect-free pipe using the same setting before testing on the pipe with a defect. A total of four tests were performed at room temperature, including measurements on a free pipe and a water-filled pipe without a defect, as well as on a free pipe and a water-filled pipe with a defect.

7.2.2.2 Data Processing

In the first case of plates, the data processing follows the same procedure as described in Section 4.3.3. However, in the second case of pipes, the measured signals need to be separated using a different approach based on the helical path separation algorithm proposed by Huthwaite [95]. Here, an overview of the background and this algorithm are briefly presented.

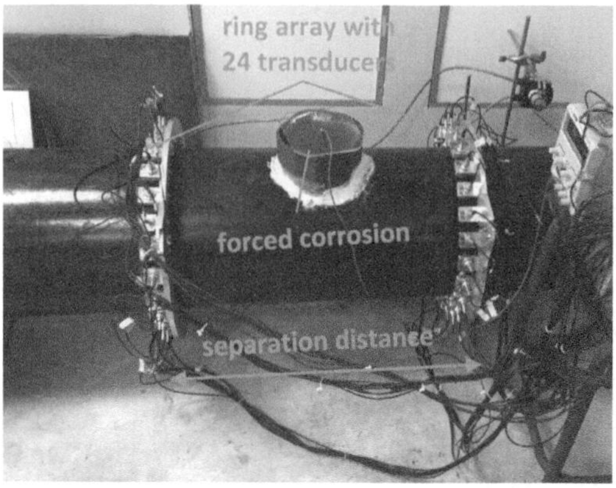

Figure 7.13 Experimental setup for measurements, including the mild steel pipe with a diameter of 12 inches and a thickness of 10.3 mm, containing a corrosion site, two ring arrays of transducers and schematic instrumentation [175].

The geometry of a pipe allows for infinite helical paths from a transmitter to a receiver due to the cyclic nature of the pipe. This is because waves can propagate multiple times around the circumferential direction of the pipe. Figure 7.14(a) shows the configuration of the pipe, with N transducers in both the transmitter array and the receiver array. Assuming the curvature of a pipe can be neglected [23, 122, 244], the pipe can be treated as a 2D plate-like domain, where the x-axis represents the circumferential direction and the z-axis represents the axial direction, as shown in Figure 7.14(b).

The circular-symmetric nature of the pipe can be accounted for by replicating the unwrapped pipe (order 0) multiple times and extending these replications (higher orders) to infinity. The signal transmitted from T_0 and received at R_0 in the pipe can be understood as a superposition of signals received at all receivers R_{-2}, R_{-1}, R_0, R_1, and R_2 on the flattened plate, as illustrated in Figure 7.14(b). It needs to be noted that the $\pm n$th replica refers to the $\pm n$th order of the helical mode. It can be seen from Figure 7.14(b) that the separation of different wave packets, corresponding to different helical order, increases the length of the two parallel linear arrays and reduces the number of missing viewing angles. This results in a less ill-posed problem.

To separate the multiple helical signals, the helical path separation algorithm proposed by Huthwaite [95] is utilized. This algorithm allows for the separation of entire wave packets, not just the first arrival times [250], even in the presence of overlapping waves. The algorithm consists of several steps: First, the measured time traces for one source and all receivers are replicated $2n$ times (from $-n$th to $+n$th order) in order to expand the aperture of the two parallel arrays and extract the information contained in higher-order modes. Second, the replicated signals containing the wavefront of interest are back-propagated to the transmitter location using the known receiver locations, the associated propagation distances, and the dispersion curves of

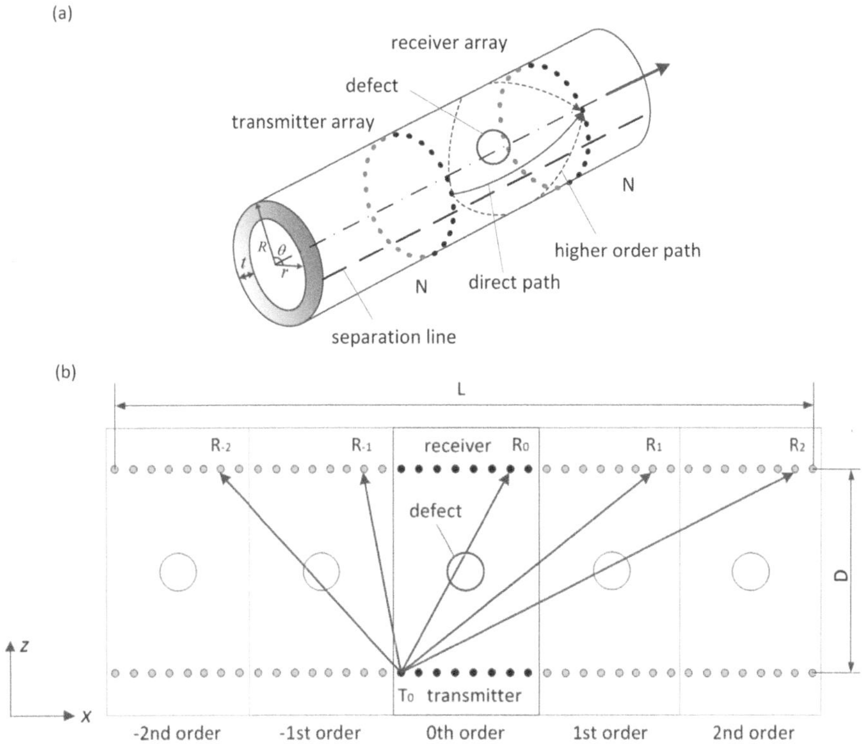

Figure 7.14 (a) 3D physical model showing the configuration of transmitter and receiver arrays and higher order helical paths, and (b) order 0 is the unwrapped section of pipe and other orders are added to both directions along the x-axis to replicate the cyclic characteristic of the pipe. In the pipe, the measured signals at R_0 correspond to a superposition of signals from each path in the flattened plate. The arrows demonstrate the various propagation paths between a transmitter and a receiver [175].

the steel plate. After the back-propagation step, the wavefront of interest is nearly uniform for all receivers in the time domain and then is windowed to remove the unwanted signals. A low pass filter is applied to all receivers in the Fourier domain, and the filtered wavefront is forward propagated back to the original locations of the receivers. Finally, the wavefront is shifted to all the non-zeroth orders and then subtracted from all the replicated signals, effectively removing the unwanted signals. This process is iterated using the obtained signals as the new original signals and adjusting the low pass filter in each iteration. The helical path separation algorithm has been validated through simulations and experiments on empty pipes with defects [95, 207].

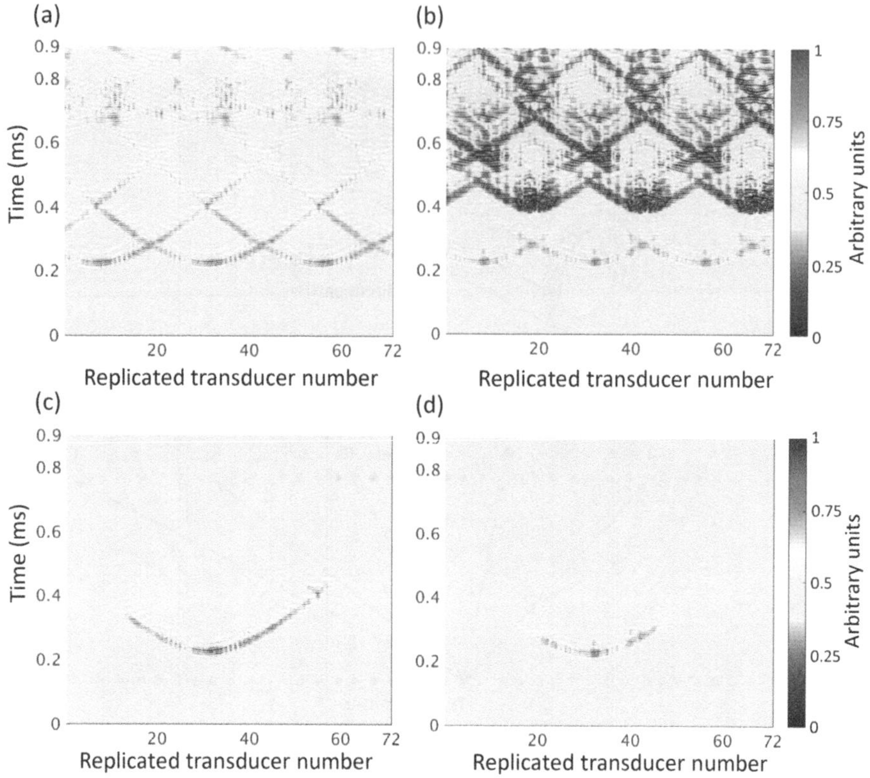

Figure 7.15 Separation of the experimental time traces for one iteration. The measured time traces are replicated two times (from $-1th$ to $+1$th order) in (a) the empty pipe and (b) the water-filled pipe; (c) and (d) the separated signals extracted by the separation algorithm from (a) the empty pipe and (b) the water-filled pipe, respectively [175].

The separation of the experimental time traces of the water-filled pipe and the empty pipe is presented in Figure 7.15. The measured time traces from one source and all receivers are replicated several times to consider different orders. Figures 7.15(a) and (b) show the replicated signals two times (from $-1th$ to $+1th$ order) obtained from a free pipe and a water-filled pipe without the defect. The wavefronts of the A_0 mode obtained from the free pipe are clearly observed. In contrast, there is less energy in the signals obtained from the water-filled pipe, especially in higher orders, which is due to higher attenuation with increasing propagation path. Additionally, it can be seen from Figure 7.15(b) that very strong signals appear after 0.4 ms, which are caused by the direct water waves resulting from the leakage of the A_0 mode.

Then, the replicated signals are separated using the helical separation algorithm [95]. The separated time traces are presented in Figures 7.15 (c) and (d). It further confirms that water loading has a significant impact on the A_0 mode, particularly in terms of amplitude. The higher-order signals exhibit more energy leakage to the water medium, resulting in a lower SNR due to the constant noise level. In Figure 7.15(d), the water waves are effectively removed after the separation algorithm and basically do not interfere with the A_0 mode. It needs to be mentioned that the separated signals contain less information (Figures 7.15(c) and (d)) compared to the corresponding parts in the replicated signals (Figures 7.15(a) and (b)). This is a result of the window applied in the receiver direction. The smoothing window is utilized to minimize spectral leakage that may occur with the discrete Fourier transform of the receiver signals [95].

The separated signals are calibrated before being used in the inversion, following a similar method described in Section 4.3.3. Subsequently, the adaptive thresholding regularization method is applied based on the results obtained from the standard limited view FWI algorithm. The regularized velocity map is then converted into the thickness map using the dispersion curves for a steel plate with and without water loading, as shown in Figure 1.2. In this analysis, the curvature of the pipe is disregarded, in accordance with [23, 117, 122].

7.2.2.3 Scattering from transducers

In the imaging algorithm, the forward modeling assumes that the incident wavefield is generated by a point excitation transducer and the scattering from the defect is produced by this incident wavefield. However, when using a contact transducer array, the incident wave can interact with other transducers, leading to the generation of secondary incident wavefields. These secondary wavefields can produce unwanted scattering components from the defect, which may affect the reconstruction accuracy of the defect [18]. Therefore, it is crucial to assess the scattering from transducers and understand its impact on the thickness reconstruction.

FE simulations and experiments were conducted to study the scattering resulting from a single transducer attached to the steel plate. The interaction between the wave generated by the transducer and its neighboring transducer in the designed array was modeled. The setup is shown in Figure 7.16(a). The scattering transducer is positioned at the center of a 300 mm diameter array, while the source transducer is located 35 mm away from the center.

The cross-section of the scattering transducer attached to the steel plate used in the FE model is shown in Figure 7.16(b), and the material properties are given in Table 4.1 and Table 7.1. The transducer consists of an aluminum cylinder, an alumina front layer, and a stack of PZT discs. The contact between the transducer and the plate was modeled using a thin coupling layer (density = 1100 kg/m^3 and acoustic impedance = 1.84×10^6 kg/m^{2}s). The model was meshed with cubic eight-node elements with a size of 1 mm, and smaller elements were used for the coupling and the front layer. The A_0 mode was excited by an out-of-plane force at the plate surface, and out-of-plane displacements were measured at 13 monitor points equally spaced along a half-circular array, as shown in Figure 7.16(a). The excitation signal was a 5-cycle Hanning windowed tone-burst at the central frequency of 60 kHz, which was the same as the one used in the thickness reconstruction. The scattered wavefield

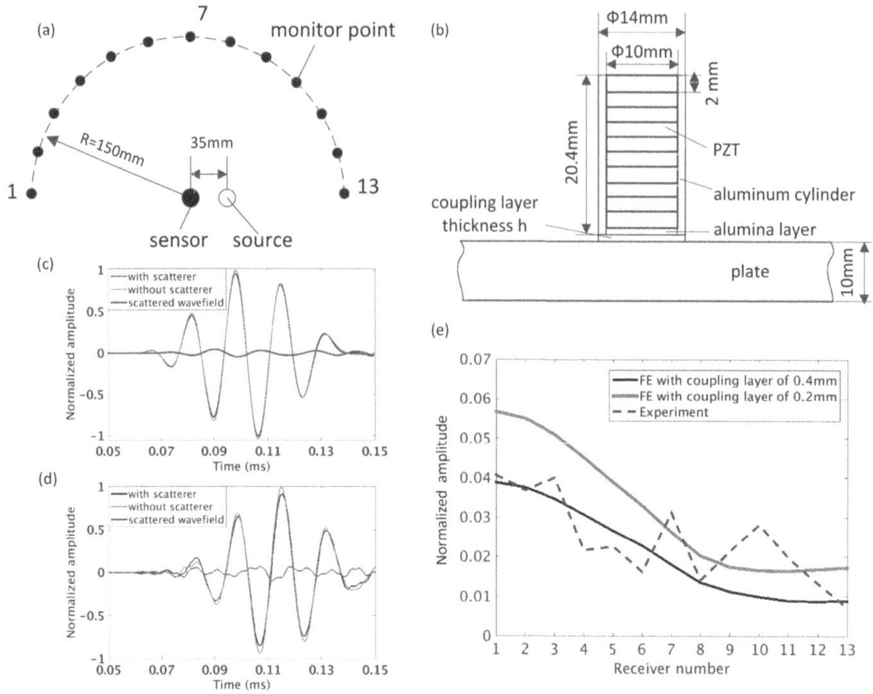

Figure 7.16 (a) Schematic of measurement for the scattering transducer positioned in the center of the array, utilizing symmetry with one source and 13 monitor points, (b) cross-section of the modeled transducer, the scattering transducer, in FE model, (c) scattered wavefield obtained by subtracting the incident wavefield of the FE simulation with and without the scatterer at measured point 1, (d) scattered wavefield obtained from the experiment at measured point 1, and (e) normalized amplitude variation with respect to the measured points in the simulation and the experiment [173].

was obtained by subtracting the incident wavefield of the model with and without the scatterer, as shown in Figure 7.16(c).

The experiments were performed on the steel plate described in Section 7.2.2.1, and two transducers from the array were used for the measurements. The scattering transducer was attached to the plate by adding weight to it. The A_0 mode was excited using the Tiepie Handyscope HS3 and the out-of-plane displacements were measured using a Polytec OFV-505 laser vibrometer. The scattered wavefield was extracted in the same manner as in FE modeling, as shown in Figure 7.16(d).

Figure 7.16(c) shows the amplitude of the scattered wavefield from the transducer at 60 kHz at different monitoring positions. The amplitude is normalized by the incident wave amplitude with the same propagation distance. Two assumptions for the thickness of the coupling layer in the FE model were used for comparison: 0.2 mm

Table 7.1
Material properties of Alumina and PZT [173]

Material	Young's modulus (GPa)	Poisson's ratio	Density (kg/m^3)
Alumina	300	0.21	3860
PZT	50	0.35	7750

and 0.4 mm. It can be seen that overall the scattered wavefield is weak, with the amplitude decreasing from the transmission part (closer to monitor point 1) to the reflection part (closer to monitor point 13). Additionally, a thinner coupling layer leads to stronger scattering. This is because with a thinner layer, more energy from the wave propagating in the plate is transmitted to the couplant [107], resulting in more energy being converted into scattering. The experimental results show a similar trend to the predictions from the FE simulation, although some variations are observed due to noise.

7.2.2.4 Scattering effect on the thickness reconstruction

The effect of transducer scattering on the thickness reconstruction of plates was investigated using FE modeling. A Hann-shaped defect with a surface diameter of 120 mm and a depth of 3 mm was modeled at the center of a steel plate with dimensions of 1100 mm$\times$1100 mm$\times$10 mm, as previously described in Section 5.3.1. The simulation involved a 700 mm diameter array with 64 transducers. Two cases were considered: one with artificial transducers (waves generated on the nodal position) and the other with real transducers as described in Section 7.2.2.3. The same input signal and reconstruction method described earlier were applied in the simulation.

The cross sections of the reconstructed thickness map are shown in Figure 7.17. Two different thicknesses of coupling layers, 0.2 mm and 0.4 mm, were modeled. It can be seen that the reconstructed thickness profiles obtained from both the artificial and real transducers are nearly identical, indicating that the scattering from the transducers has minimal impact on the accuracy of the thickness reconstruction.

7.2.3 EXPERIMENTAL SIMULATION OF CORROSION DAMAGE

7.2.3.1 Forced electrochemical corrosion

While unforced corrosion is more representative of real-world cases, forced corrosion, which is driven by an applied electric current, is used to accelerate corrosion damage in a relatively short time. In the case of steel corrosion, this reaction is dominated by the two-electron oxidation of elemental iron, $Fe \rightarrow Fe^{2+} + 2e^{-}$. In the forced corrosion experiments conducted in this chapter, the mild steel plate or pipe was used as the anode, where the oxidation of iron takes place. A steel thin disk was inserted into the chamber, serving as the cathode (as shown in Figure 7.18(a)).

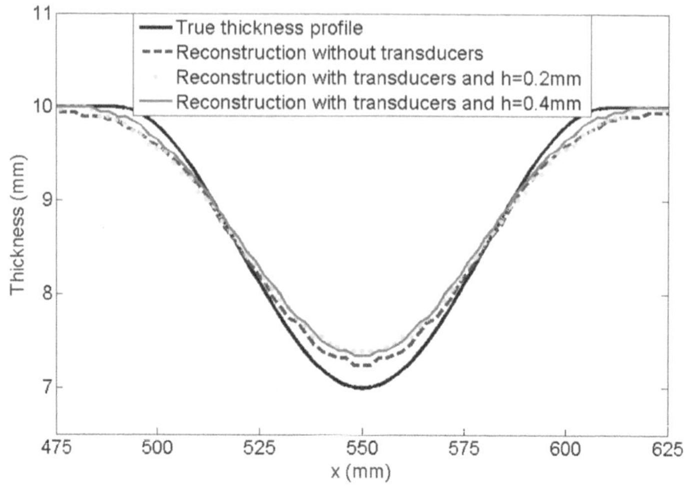

Figure 7.17 Thickness profiles across the reconstructions of a Hann-shaped defect with a surface diameter of 120 mm and a depth of 3 mm from the FE model in the plate without and with an array of modeled transducers. *h* is the thickness of the coupling layer [173].

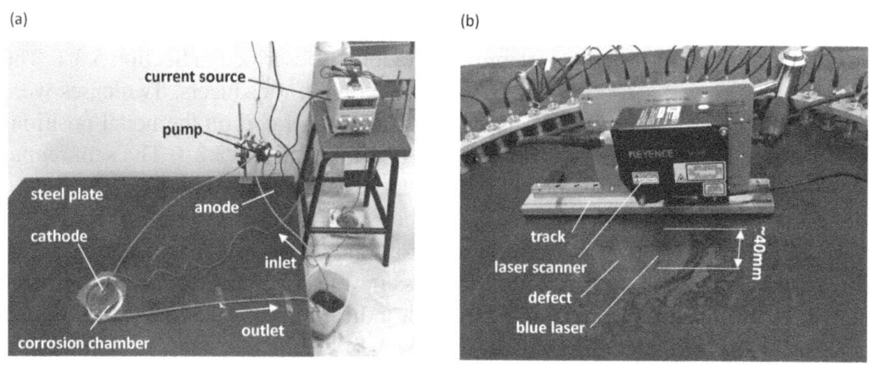

Figure 7.18 (a) The experimental setup of the forced corrosion and (b) a laser profile scan of the corrosion defect in a 10 mm steel plate [173].

This setup allows for controlled and accelerated corrosion to occur, facilitating the collection of results within a relatively short time frame.

Figure 7.18(a) shows the experimental setup for forced electrochemical corrosion on the mild steel plate. A corrosion chamber with a diameter of 130 mm is attached to the surface of the steel plate at the location of the artificial defect. During the corrosion process, a brine electrolyte is circulated through the chamber using a pump (Verderflex M025, Verder Ltd., Castleford, UK). This circulation helps prevent the precipitation of corrosion products, such as iron oxide, and maintains a steady corrosion rate. Additionally, it helps prevent the further oxidation of iron oxide into iron hydroxide, which is insoluble and acts as a protective layer against further corrosion. Moreover, circulating the electrolyte helps minimize temperature changes during the electrochemical corrosion process, as significant temperature increases can lead to reductions in ultrasonic velocities [205]. In the forced corrosion experiment, the electric current is supplied by a constant current source (IPS 303DD, ISO TECH, UK).

Faraday's law of electrolysis is used to predict the average wall loss caused by the applied current in the forced corrosion experiment. According to this law, the average wall loss ΔT_h of the steel plate can be calculated by

$$\Delta T_h = -\frac{M_{Fe} It}{2FA_c \rho_{Fe}}, \tag{7.3}$$

where M_{Fe} is molar mass of steel (0.055845 mol/kg), I being the applied current, and t is the elapsed time. F is the Faraday's constant (96485.3329), A_c is the area of the corrosion surface (diameter of 0.13 m), and ρ_{Fe} is the density of steel (7850 $\mathrm{kg/m^3}$).

In this chapter, a constant current of 3 A was used in the forced corrosion experiment. The relationship between the wall loss of the steel plate and time can be expressed by

$$\Delta T_h = -8.3324 \times 10^{-9} t. \tag{7.4}$$

7.2.3.2 Thickness measurement using laser profilometry

In this chapter, a blue laser scanner (LJ-V7080, KEYENCE, Japan) was utilized to measure the wall losses and profiles of the controlled corrosion damage. An example of laser scanning is shown in Figure 7.18(b). The laser scanner offers a repeatability precision of 10 μm along the width of the laser beam and 0.5 μm along the thickness of the plate. The acquired data is processed using the software provided by the manufacturer. It is important to note that the laser scanner has a maximum measuring range of approximately 40 mm in width. Therefore, for larger sizes of corrosion damage, multiple measurements are taken and then merged to obtain a comprehensive assessment.

Before conducting measurements with the laser scanner, the forced corrosion setup is dismantled, and special attention is given to meticulously clean the corrosion surface to remove any residual corrosion products. This cleaning process ensures that the laser scanner can accurately measure the corrosion profile without interference from any remaining debris or contaminants.

7.2.4 RESULTS AND DISCUSSIONS

The focus of this chapter is on evaluating the performance of the UGWT system for corrosion monitoring of both the plate and the pipe. In the case of the plate, the first step is to compare the reconstructed depths of the corrosion defect at different stages using the FWI algorithm with the values predicted by Faraday's law. Subsequently, a detailed comparison is conducted between the thickness maps of the plate obtained from the FWI algorithm and those obtained using laser profilometry at two different corrosion periods. This comparison provides insights into the effectiveness of the FWI algorithm in accurately capturing corrosion-induced thickness changes over time.

Moving on to the corrosion monitoring of the pipe, the study focuses on the thickness reconstruction of an irregularly shaped defect in both the empty pipe and the water-filled pipe. The goal is to examine the performance of the UGWT system in accurately identifying and characterizing localized corrosion damages in different pipe conditions. By comparing the reconstructed thickness maps of the defect in the empty pipe and the water-filled pipe, the effectiveness of the UGWT system in detecting and assessing corrosion in different environments can be evaluated.

7.2.4.1 Corrosion monitoring of plate

7.2.4.1.1 Wall loss monitoring

In this work, measurements of the corrosion damage induced by forced corrosion were conducted over a period of approximately 110 hours. Based on the relationship described by Equation (7.4), an average thickness loss of 1 mm can be achieved after approximately 36 hours of corrosion. This information provides an estimate of the corrosion progression rate and helps in understanding the time scale required to induce significant thickness loss in the experimental setup.

To compare the average reconstruction obtained through FWI with the average wall loss calculated using Faraday's law, an alternative metric for quantifying the average reconstructed thickness from FWI is to take the mean value of all reconstructed thickness within the upper limit of the minimum reconstructed thickness $(1+5\%)T_{min}$ [90]. The average reconstructed thickness $\overline{T}$ can be given by

$$\overline{T} = \sum_{n=1}^{N} \frac{T(x_n, y_n)}{N} \quad when \quad T(x_n, y_n) \leq (1+5\%) \times T_{min}, \tag{7.5}$$

where $T(x_n, y_n)$ is the reconstructed thickness at the coordinates x_n and y_n; T_{min} represents the minimum reconstructed thickness, and N denotes the number of reconstructed thickness values within the range of $(1+5\%)T_{min}$.

Figure 7.19 shows the average reconstructed thicknesses of defects at different stages during the corrosion process obtained through the FWI algorithm, along with analytical predictions based on Faraday's law. The results demonstrate a high level of agreement between the average reconstructed thicknesses and the analytical predictions during the continuous corrosion.

7.2.4.1.2 Thickness reconstruction

In Chapter 5, it was demonstrated that the FWI algorithm achieves a resolution of approximately 1.5-2λ for most of the defect depths in the experimental setup.

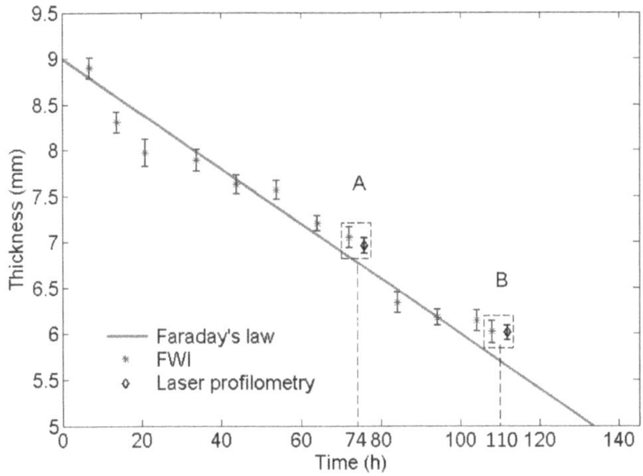

Figure 7.19 The comparison of the average reconstructed thickness of the defect obtained by FWI at 60 kHz and the predictions by Faraday's law in different corrosion periods of a mild steel plate. The upper and lower limits of the measured thickness of defects A and B at corrosion time of 74 and 110 hours obtained by FWI and laser profilometry are shown in the rectangular boxes, respectively. The results of laser profilometry will be explained in Section 7.2.4.1.2 [173].

Furthermore, the shape of the defect profile was found to have an impact on the resolvability. In this chapter, the effect of different surface and bottom diameters, as well as the depths of flat-bottom defects in various corrosion progressions, on the reconstruction accuracy is verified.

Figure 7.19 presents the results of corrosion defects A and B at the corrosion times of 74 and 110 hours, obtained from both the UGWT system and a laser profile scan. The reconstructed thickness profiles are depicted within two dotted boxes, and the wall losses are represented by error bars, indicating the upper and lower limits of the measured thickness. It is worth noting that the reconstructions from the UGWT system and the laser profile measurements exhibit excellent agreement, even surpassing the predictions from the analytical model.

Figure 7.20(a) shows the profile measured by laser scans after 74 hours of corrosion. As mentioned earlier, due to the limitation of the laser scanner, the defect profile needs to be scanned in separate strips, resulting in a slightly discontinuous spliced profile. However, it is clear from the figure that the quality of the interior of the resolvable area remains intact. The defect exhibits a regular shape with relatively smooth variations in thickness. The maximum extent of the defect is approximately 130 mm ($\simeq 3.9\lambda$ at 60 kHz), and the maximum depth is around 3 mm. In this study, sequential frequencies of 35 kHz, 46 kHz, and 60 kHz were utilized for the inversion.

Figures 7.20(b) and (c) present the thickness reconstruction results and a zoomed image of the defect at 60 kHz, respectively. It can be observed that the overall shape

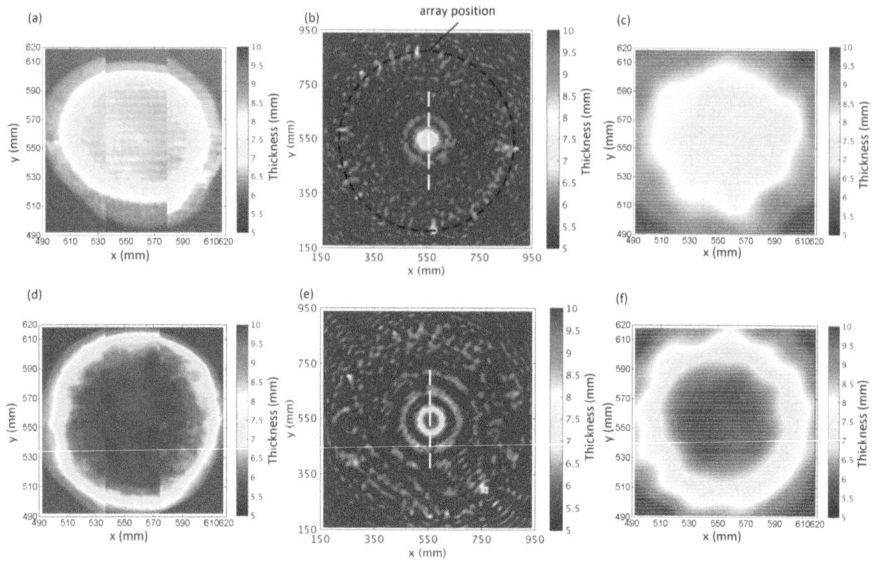

Figure 7.20 Thickness mapping of the corrosion defects A and B in the steel plate. (a) Laser-measured profile of defect A after post-processing, (b) thickness reconstruction from FWI at 60kHz, and (c) a zoomed-in view of defect A from (b). The dashed curve presents the position of the array. (d) Laser-measured profile of defect B after post-processing, (e) thickness reconstruction from FWI at 60 kHz, and (f) a zoomed picture of defect B from (e) [173].

of the reconstructed defect A closely approximates the measured corrosion profile. However, there is a slight deformation in the shape of the defect, and the contrast of the defect is slightly underestimated in the experimental reconstruction. This is attributed to the influence of different scattering behaviors between guided waves with thickness variation and acoustic waves with velocity variation [91]. Some artifacts are noticeable in the reconstructed image, particularly at the positions of the transducer array and around the defect. These artifacts may be caused by factors such as environmental noise in the experiment, undersampling of the wavefield at the array position [218], and the scattering from the array of transducers. The artifacts around the defects can mainly be attributed to the presence of the chamber wall and the silicon used in the bonding of the chamber wall to the plate, which were not considered in the reconstruction algorithm.

The thickness profiles of a deeper defect B after 110 hours of corrosion are shown in Figures 7.20(d) and (f). It is clear that the reconstruction can still capture most of the features of the defect, although the discrepancy compared to the measured thickness profile is larger than in the previous case, as the error of the reconstruction increases with the depth of the defect [207].

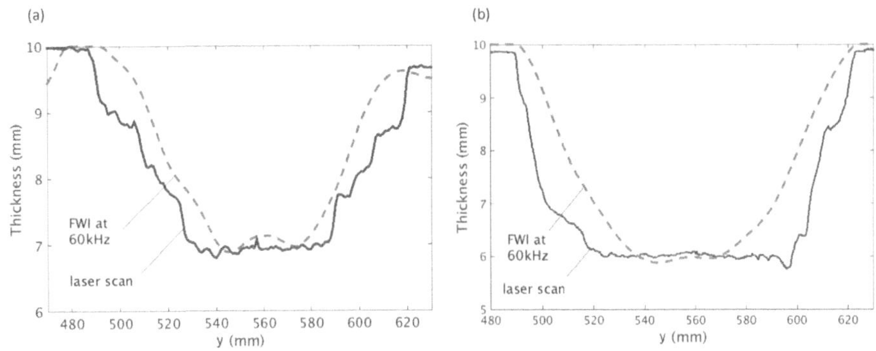

Figure 7.21 Cross sections of FWI reconstruction and laser measurement for (a) defect A and (b) defect B along the vertical line in the plate [173].

The cross sections of the thickness profile along the vertical direction (dashed lines in Figure 7.20) for both defects are plotted in Figure 7.21. It can be seen that the location of the defect and the minimum thickness have been successfully identified for both cases. It can also be noted that the reconstructed defects are slightly narrower than the measurements from the laser scanner, which is believed to be caused by the difference in scattering between the acoustic and elastic models [91].

7.2.4.2 Corrosion mapping of pipe

The thickness reconstruction of an irregularly shaped defect in the empty pipe allows for the assessment of achievable reconstruction accuracy. To analyze the impact of attenuation due to the presence of water loading on the reconstruction accuracy, the water-filled pipe is investigated.

7.2.4.2.1 Empty pipes

The profile of the irregularly shaped defect induced by forced corrosion is measured by a laser scanner, as shown in Figure 7.22(a). This defect has the largest width of around 150 mm ($\simeq 4.4\lambda$ at 60 kHz) and a depth of around 4 mm. To analyze the benefits of using replications (higher-order modes), Figure 7.22 compares the thickness reconstructions of the empty pipe obtained by considering four replicas (Figure 7.22(b)) and two replicas (Figure 7.22(c)). As anticipated, the larger aperture results in a better quality reconstruction in terms of defect shape and depth estimation. However, both reconstructions slightly underestimate the size of the defect. The discrepancy is mainly attributed to the limited view configuration of the UGWT system. Since these data are measured from the pipe with two ring arrays, some information is missed out on certain viewing angles even with the arrays over four replicas. The remaining discrepancy is due to the difference between acoustic and elastic scattering, as the acoustic reconstruction algorithm is employed to reconstruct the elastic scattering data from the experiment [92, 207].

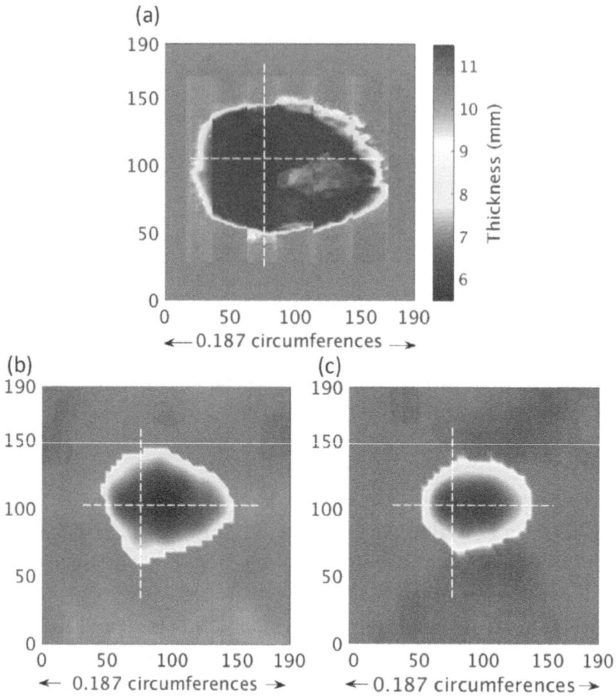

Figure 7.22 Thickness mapping of an irregularly shaped defect in the empty pipe. (a) The profile of the defect was measured by a laser scanner, (b) the FWI reconstruction with four replicas, and (c) the FWI reconstruction with two replicas at 60 kHz [175] (unit = mm).

The cross sections of Figure 7.22 along the horizontal and vertical lines are presented in Figures 7.23(a) and (b). The profiles of the defect with four and two replicas are narrower compared to the measurement from the laser scanner, as explained earlier. However, it is evident that the reconstruction with four replicas provides slightly better estimations of the defect profile and depth compared to the reconstruction with two replicas, primarily due to a larger range of angles available for the inversion. The maximum depths of the defect with four and two replicas are underestimated by around 0.2 mm and 0.7 mm, respectively.

7.2.4.2.2 Water-filled pipes

In Figure 7.24, the defect profile measured by the laser scan is compared with the reconstruction thickness of the water-filled pipe. In this case, only two replicas are used in the inversion due to the difficulty in identifying the attenuated signals that gradually leak energy into the water with increasing helical path, especially in higher-order paths (e.g., ±2th order). Compared to the results of the empty pipe shown in Figure 7.22(c), the defect size is slightly smaller and its shape is distorted.

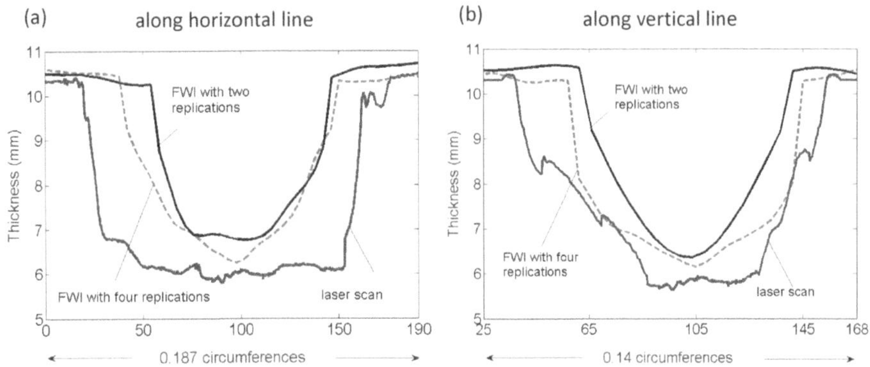

Figure 7.23 Cross sections of the FWI reconstructions of an irregular defect and laser measurement along the line marked in Figure 7.22: (a) the horizontal line and (b) the vertical line in the empty pipe [175] (unit = mm).

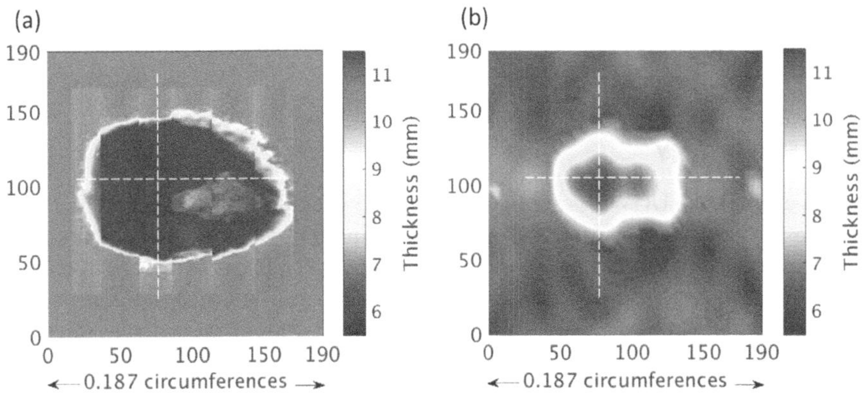

Figure 7.24 Thickness mapping of an irregularly shaped defect in the water-filled pipe. (a) The profile of the defect measured by a laser scanner (same as in Figure 7.22(a)), and (b) the FWI reconstruction obtained considering two replicas with the attenuation in the inversion model at 60 kHz [175] (unit = mm).

Additionally, more noticeable artifacts can be observed in the reconstruction, even when considering the attenuation in the regularization process. The main reason is that the effective array aperture in the water-filled pipe is smaller than that in the empty pipe due to the energy leakage to the water medium, as shown in Figures 7.15(c) and (d). This reduced aperture affects the accuracy of the thickness reconstruction.

Figures 7.25(a) and (b) show the cross sections of Figure 7.24 along the horizontal and vertical lines, respectively. The thickness reconstructions of the deepest

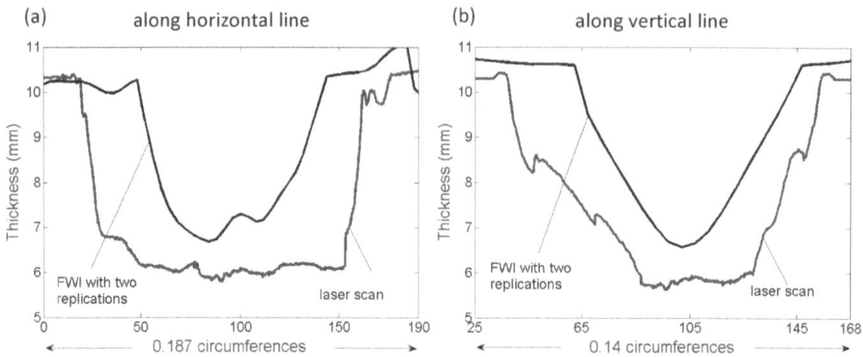

Figure 7.25 Cross sections of the thickness reconstruction from Figure 7.24 along (a) the horizontal line and (b) the vertical line in the water-filled pipe [175] (unit = mm).

points, taking into account attenuation, exhibit a slight underestimation with a thickness error of approximately 0.7 mm. Overall, the estimates of the maximum defect depth in both the empty pipe and the water-filled pipe, considering attenuation in the inversion model, align well with each other.

7.3 SUMMARY

In this chapter, the UGWT is applied to plate-like structures filled with liquid, utilizing the FWI method. The focus is on investigating the effect of attenuation caused by the presence of liquid loading on the accuracy of thickness reconstruction using the S_0 mode. Through simulations considering attenuation at different frequencies, a range of defect widths and depths on the water tank floor is examined to evaluate the suitability of the S_0 mode for precise defect reconstruction. The results indicate that the S_0 mode exhibits improved sensitivity at higher frequencies, enabling accurate reconstruction of defects. The impact of attenuation on thickness reconstructions in liquid-loaded tank floors is generally small when the attenuation is low. However, as the attenuation increases, particularly in the presence of highly attenuating liquids, it becomes crucial to consider attenuation for more accurate reconstructions. Experimental results on a water tank floor with an irregularly shaped defect validate the effectiveness of the UGWT approach. The reconstructed thickness of the defect, considering attenuation in the inversion model at a frequency of 150 kHz, closely matches its actual depth.

Based on the previous work, a UGWT system has been developed for efficient and accurate corrosion monitoring of plate-like and pipe structures. The system utilizes an array of piezoelectric transducers designed to generate and receive the A_0 Lamb wave at low frequencies. It has been observed that the scattering from the selected transducers has minimal impact on the accuracy of thickness reconstruction.

In the forced corrosion experiment of the plate, the average wall losses that were reconstructed by FWI agreed well with the analytical predictions obtained by Faraday's law. The reconstructed thickness maps of the plate at two different stages matched well with the measurements from a laser scanner. In the forced corrosion experiment conducted on the plate, the average wall losses reconstructed using the FWI algorithm closely matched the analytical predictions derived from Faraday's law. The reconstructed thickness maps of the plate at different stages were also in good agreement with the measurements obtained from a laser scanner.

For the pipe with a limited view configuration, an adaptive regularization method and the information carried by higher-order modes obtained through the helical path separation technique were employed to enhance the reconstruction accuracy. In general, the results show that the maximum depth estimation of the defect in a water-filled pipe with the consideration of attenuation aligns well with the estimation obtained from an empty pipe. Overall, these findings highlight the effectiveness of the UGWT system in accurately assessing corrosion in both plate-like and pipe structures, offering valuable insights for corrosion monitoring applications.

7.4 EXERCISES

1. Please draw the dispersion curves of Lamb waves in the aluminum plate with and without water loading.
2. Please analyze how attenuation is considered in the frequency domain and time domain full waveform inversion, and what are the advantages of frequency domain full waveform inversion with the consideration of attention?
3. Please analyze the reasons for choosing A_0 mode or S_0 mode.
4. Please describe the effect of attenuation as a function of defect type, depth, and width.
5. Please analyze the relationship between attenuation and reconstruction accuracy.
6. Please combine with Figure 7.7 to analyze the impact of considering attenuation and not considering attenuation on reconstruction accuracy and the causes of errors.
7. Please analyze the causes of the error in Figures 7.8 and 7.9.
8. Please analyze the reasons for the error between the experimental results in Figure 7.10 and the theoretical values.
9. What parts does the UGWT system generally consist of?
10. What are the differences between the self-designed low-frequency sensor used in this chapter and the commercially purchased low-frequency sensor?
11. Please describe the data process of the ultrasonic guided wave in the pipe, and the difference and connection with the guided waves in the limited-view plate.
12. Please analyze the scattering from the self-designed transducer.
13. Please analyze the causes of the error between the FWI reconstruction and the result obtained by a laser scanner in Figures 7.22 and 7.24.

8 Applications of Material Characterization

8.1 QUANTITATIVE IMAGING OF YOUNG'S MODULUS IN PLATES

8.1.1 INTRODUCTION

Young's modulus is a crucial parameter for measuring material stiffness, with applications in fields like medicine and structural mechanics. It helps in understanding the condition of materials and structures, making it valuable in various industries. For example, in aerospace, identifying local stiffness loss in composite laminates due to impact damage is critical, as it can significantly impact strength. To detect such issues early, there is a need for quantitative methods to evaluate Young's modulus.

One common technique for measuring Young's modulus is the pulse-echo method, which uses the time-of-flight of waves to calculate the velocities of longitudinal and shear waves [156]. However, this method is best suited for local measurements that require manual or automated scanning for larger areas, which can be time-consuming and expensive.

Lamb waves and other guided ultrasonic waves are more appropriate because they overcome the shortcomings of local measurements and increase coverage. They are able to traverse large distances within a plate, offering insights into the characteristics of materials over wider areas. However, estimating the elastic properties of Lamb modes is complex because of their dispersive nature. Their phase velocity depends on Young's modulus and Poisson's ratio, and this information is obtained by comparing the measured dispersion curves with theoretical predictions [52, 105, 195, 272]. If one elastic constant is known, Young's modulus can be determined from certain non-dispersive modes [21, 119]. Still, these techniques assume material properties to be constant along the wave's path, making it challenging for materials with non-uniform properties.

UGWT offers a way to image spatially varied material properties by analyzing ultrasonic signals captured by a transducer array around the inspection area. Previous studies mainly focused on assessing corrosion damage, but we aim to extend UGWT to image Young's modulus variations in a plate. This involves velocity mapping through the use of FWI, a non-linear inversion technique that is performed iteratively across multiple frequencies. In the following sections, the principle of reconstructing Young's modulus from the velocity of the fundamental A_0 mode using FWI is described. This is followed by the presentation of the results of the numerical study obtained through finite element simulations and the sharing of experimental details along with the validation results.

DOI: 10.1201/9781003667926-8

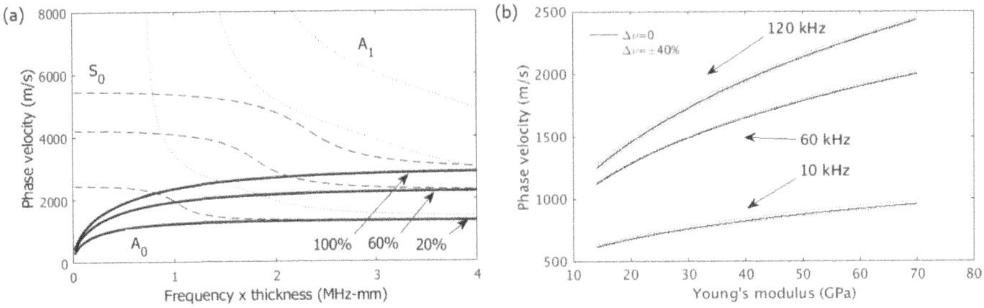

Figure 8.1 Phase velocity dispersion curves for (a) Lamb waves in 10 mm thick aluminum plate with varied Young's modulus of 20, 80, and 100% of the nominal value; (b) A_0 mode as a function of Young's modulus plotted at different frequencies and Poisson's ratio variation $\Delta\nu$ [187].

8.1.2 NUMERICAL STUDY

8.1.2.1 Young's modulus obtained from phase velocity

The velocities of the bulk longitudinal c_l and shear c_s waves are given by

$$c_l = \sqrt{\frac{E(1-\nu)}{\rho(1+\nu)(1-2\nu)}} \text{ and } c_s = \sqrt{\frac{E}{2\rho(1+\nu)}}, \tag{8.1}$$

where E is Young's modulus, ν is Poisson's ratio, and ρ is density. Therefore, the phase velocity of an ultrasonic guided wave is related to Young's modulus of the waveguide. However, the relationship between the phase velocity of ultrasonic guided waves and Young's modulus is intricate and requires solving the dispersion equations [79].

The effect of Young's modulus on the phase velocity of guided waves is explored in Figure 8.1(a). The dispersion curves for the first three Lamb modes (A_0, S_0, and A_1) are calculated for a 10 mm thick aluminum plate. Young's modulus is intentionally reduced by 40% and 80% from its nominal value, while Poisson's ratio remains constant. The results clearly demonstrate that all modes are responsive to Young's modulus, and their velocities decrease as Young's modulus decreases. This is as expected since the wavenumber, inversely proportional to velocity, is affected by Young's modulus.

Furthermore, the influence of Young's modulus and Poisson's ratio on phase velocity at lower frequencies is depicted in Figure 8.1(b). It can be seen that velocity decreases with decreasing Young's modulus, while the effect of Poisson's ratio remains relatively small, even when varied by ±40%. This finding aligns with previous studies on the impact of Poisson's ratio on Lamb modes [195, 200]. This indicates that velocity measurement is not only sensitive but also selective for capturing variations in Young's modulus within a plate.

Table 8.1
Configuration parameters used for FE modeling [187]

Model nr	Plate size (mm^2)	Array diameter (mm)	Element size (mm)	Defect width (mm)	Excitation frequency (kHz)	Nr of cycles
1	1000 x 1000	800	1	60, 120	50	5
2	500 x 500	400	0.5	30, 60	100	3

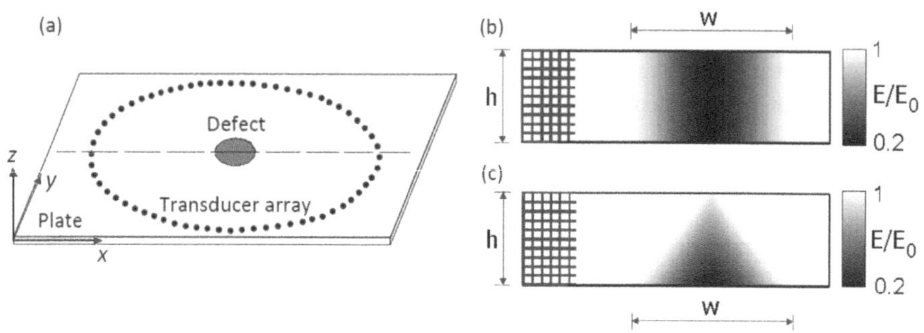

Figure 8.2 (a) Configuration of the plate setup, with a central stiffness defect and a circular transducer array surrounding it; (b) cross-section of the meshed plate with a defect of constant through-thickness normalized Young's modulus distribution [187].

8.1.2.2 Configuration of the models and defects and data processing

To evaluate the performance of the reconstruction algorithm under various Young's modulus reduction scenarios, 3D finite element (FE) simulations were conducted using the Abaqus software. The model configuration is depicted in Figure 8.2(a). It consists of a 10 mm thick aluminum plate ($E = 70$ GPa, $\nu = 0.33$, $\rho = 2700$ kg/m^3) with a Young's modulus reduction at its center. A circular array of 64 transducers was placed around the area to be imaged. Parameters such as plate size, element size, array diameter, and excitation tone-burst were adjusted to optimize simulations for different frequencies and defect sizes. The specific parameters for the two selected models are summarized in Table 8.1. Eight-node Cubic elements were used to ensure accurate modeling, with more than 30 elements per wavelength at the highest frequency. In addition, an absorbing region was included to minimize reflections from the edges. Out-of-plane displacement was applied at one surface node, and wavefields were measured by the other 63 elements. This process was repeated for all source-receiver combinations, generating a 64×63 signal matrix. The source excitation was a multi-cycle Hanning windowed tone-burst signal at a given center frequency to ensure a nearly pure A_0 mode.

Defects with desired stiffness distributions were created by adjusting the Young's modulus of elements in the plate mesh. Initially, axisymmetric Hann-shaped defects were used, situated at the plate's center. The Young's modulus variation within these defects is given by

$$E(r_c,z) = \begin{cases} E_0 - \frac{(E_0 - E_{min}(z))}{2}[1 + cos(\frac{2\pi r_c}{w(z)})] & r_c < w/2 \\ E_0 & r_c > w/2 \end{cases}, \tag{8.2}$$

where E_0 is the nominal Young's modulus, $E_{min}(z)$ is the minimum Young's modulus at the center of the defect along the thickness, $w(z)$ is the outer diameter of the defect, and r_c is the radius from the center. Various cases can be investigated by changing the parameters $E_{min}(z)$ and $w(z)$.

Two types of axisymmetric defects were considered. First, Young's modulus in the defect area is varied only along the plate plane, and its magnitude and width are considered to be constant through the plate thickness, as shown in Figure 8.10(b) with a scale of the normalized Young's modulus. This case closely resembles the simplified 2D acoustic model where all the parameters of the plate are averaged to the mid-plane of the plate. In this case, the symmetry of the defect with respect to the mid-plane of the plate prevents mode conversions from occurring when the waves interact with it, further simplifying the problem.

In real-world scenarios, the stiffness distribution throughout the thickness may not be constant. For such cases, a simplified linear through-thickness variation is considered, as shown in Figure 8.10(c). Its width is linearly reduced from w to 0 and its Young's modulus increases from E_{min} to E_0 from the bottom to the top surface. Clearly, it is not possible to model this defect for the acoustic case, and an average through-thickness Young's modulus should be considered.

$$E_{av} = \frac{\int_0^h E(r_c,z)dz}{h}, \tag{8.3}$$

where h is the thickness of the plate. Also, here mode conversions are expected to be small as a smooth stiffness variation is being considered. Finally, irregularly shaped stiffness variation along the plate plane is considered with variation constant along the thickness.

In the data processing stage, frequency domain data necessary for the inversion were obtained by applying a fast Fourier transform to the time-traces from each source-receiver pair. A calibration factor was calculated by comparing transformed data from homogeneous FE and FD models, as described in Section 4.3.3. The FWI calculation domain consisted of a grid with a uniform velocity distribution as an initial guess. The inversion process was implemented sequentially, starting with lower frequencies. Each frequency group underwent up to 30 iterations for convergence, with a criterion for a relative change in the velocity field. Gaussian smoothing was applied to the perturbation model to minimize high-frequency artifacts in the gradient. After inversion, the reconstructed velocity maps were converted to Young's modulus maps using the dispersion relationship between Young's modulus, frequency, and phase velocity.

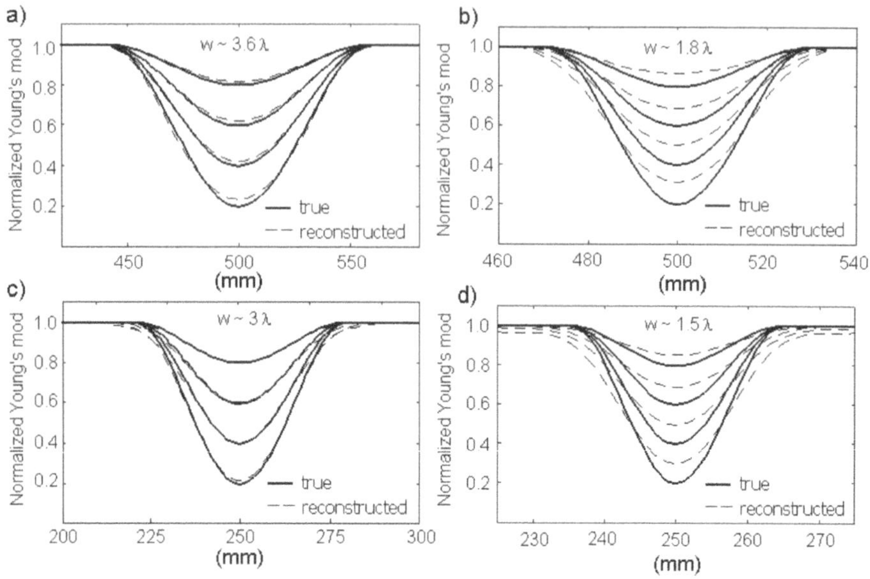

Figure 8.3 Cross sections of reconstructions of a single central defect with constant through-thickness Young's modulus variation. (a) Defect width w = 120 mm at 60 kHz; (b) w = 60 mm at 60 kHz; (c) w = 60 mm at 120 kHz; (d) w = 30 mm at 120 kHz [187].

8.1.2.3 Young's modulus reconstruction results

In the first modeling case, a constant through-thickness Young's modulus was used, with Hann-shaped defects along the plate's center within a 64-element circular array. Cross-sections of the reconstructions are presented in Figures 8.3(a), (b), (c), and (d) for Young's modulus drops of 20%, 40%, 60%, and 80% from the nominal value. The defect widths were 120 mm and 60 mm at 60 kHz and 60 mm and 30 mm at 120 kHz. Young's modulus is normalized by the nominal value.

Figures 8.3(a) and (c) show that the reconstructions are highly accurate for larger defects at both frequencies. However, for smaller defects (where the width is approximately 1.5λ), the reconstructions tend to underestimate the true Young's modulus drop and slightly overestimate the width of the defect. The absolute reconstruction error appears to increase with Young's modulus drop, with slightly better performance observed at the higher frequency.

In the second modeling case, similar defects with linear changes in the linear thickness parameter were considered. The width of the defect increased from 0 to its prescribed value (w) from the top surface to the bottom surface, and Young's modulus decreased by 40% and 80% from the nominal value. Figures 8.4(a) and (c) indicate that for the larger defect, the reconstruction slightly overestimates the true averaged through-thickness value. For smaller defects, the peak drop is reconstructed more accurately.

The final case involves irregularly shaped stiffness defects. The stiffness is assumed to be constant throughout the thickness. The reconstruction at 60 kHz, as

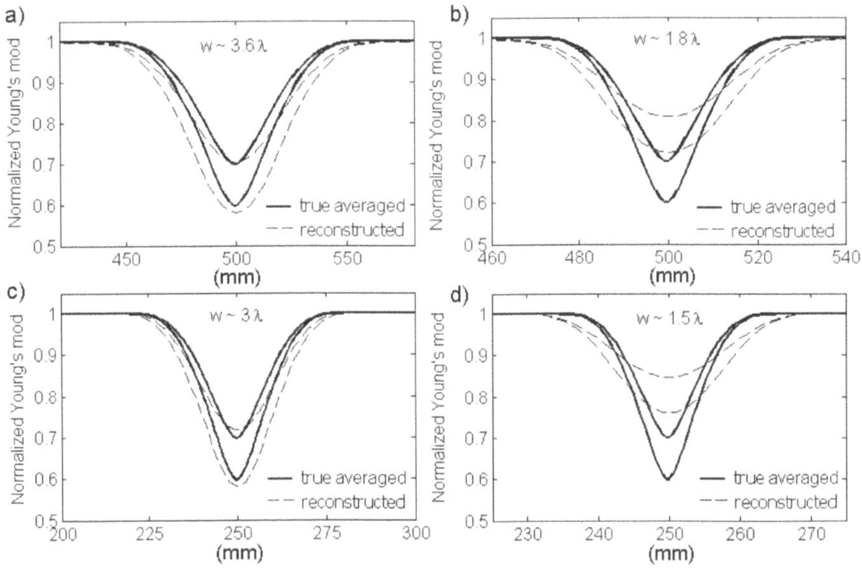

Figure 8.4 Cross sections of reconstructions of a single central defect with linear through-thickness Young's modulus variation for $E_{min}/E_0 = 0.2, 0.6$. (a) Defect width w = 120 mm at 60 kHz; (b) w = 60 mm at 60 kHz; (c) w = 60 mm at 120 kHz; (d) w = 30 mm at 120 kHz [187].

shown in Figure 8.5(b), accurately captures all the details. The cross sections of the original and reconstructed Young's modulus distribution are compared in Figure 8.5(c), demonstrating a very close agreement between the two, with comparable errors in the minima. This demonstrates the robust performance of FWI in irregularly shaped corrosion defects, as described in Chapter 4.

Simulation results revealed a decrease in resolution and accuracy as the defect size approached the limit of $\lambda/\sqrt{2}$ for transmitted waves, particularly evident in the nonconstant variation of Young's modulus. This decline is attributed to the limitations of the 2D acoustic approximation in capturing the complete physics compared to a more realistic 3D wave propagation model. Huthwaite's study [90] supports this, demonstrating the degradation in reconstruction accuracy when comparing simulated scattering from small velocity variations using acoustic and elastic models in a limited scattering region.

8.1.3 EXPERIMENTAL RESULTS

8.1.3.1 Setup and measurements

In the experimental setup, the desired variation in Young's modulus of the metal plate was achieved by heating it. The setup, which illustrates the temperature field

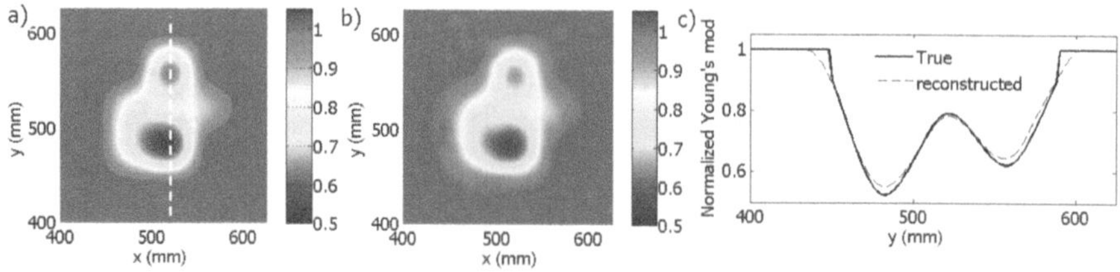

Figure 8.5 (Color online) Normalized Young's modulus of an irregularly shaped defect. (a) Original map; (b) reconstructed map from FE modeling at 60 kHz; and (c) vertical cross-section through $x = 510mm$ [187].

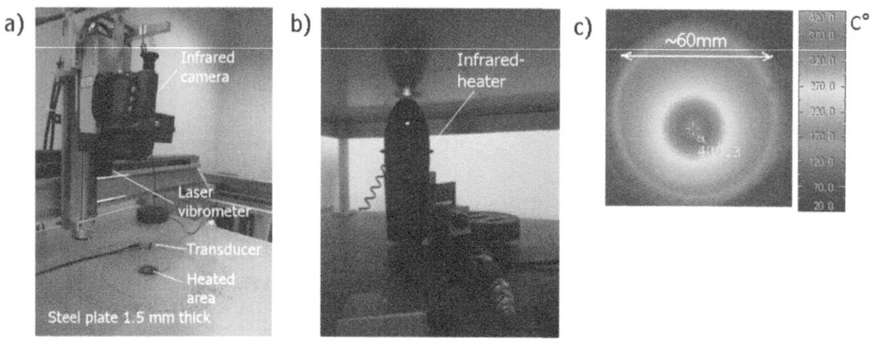

Figure 8.6 (a) Experimental setup to measure ultrasound propagation in a heated plate; (b) plate spot heating with an infrared heater below the plate; (c) typical temperature measurement of the heated area by the infrared camera [187].

and the measurement of the propagation of the ultrasonic wave on the plate, is shown in Figure 8.6. It is well-known that for most metallic materials, Young's modulus decreases as the temperature increases.

A steel plate (SUS304) measuring 1000 mm × 1000 mm × 1.5 mm was used. The plate was supported at its edges with a frame and heated using an infrared heater (4185 IR Spot Heater) positioned underneath the plate, as shown in Figure 8.6(b). The heater mainly consists of a power-controlled tubular quartz infrared lamp emitting infrared energy, which is then focused and directed towards the desired target area. This allows non-contact heating with temperatures up to 800°C.

Before heating, a small circular area to be heated was coated with a thin layer of black heat-resistant paint (Nippon paint Pylox) on both sides of the plate. This coating helps slow the rapid temperature drop as a result of heat conduction in metallic materials and minimizes the temperature gradient through the thickness. Previous studies have shown that this thin coating has a negligible effect on wave propagation in plates [94].

Table 8.2
Material properties of a steel plate at room temperature [187]

Young's modulus (GPa)	Poisson's ratio	Density (kg/m^3)
184	0.2685	8051

The plate was heated for approximately 30 minutes to reach a stable working temperature. Temperature monitoring was carried out using an infrared camera (Thermo Tracer TH9100 Pro) and cross-checked with a noncontact laser infrared temperature gun, showing good agreement between the two devices. The temperature distribution in the heated area, measured by the infrared camera, is depicted in Figure 8.6(c), with a peak value around 400°C, smoothly decreasing away from the heating spot. The peak temperature value was monitored during ultrasound measurements, with an error of no more than ±10°C. It is expected that the temperature remains nearly constant throughout the thickness as a result of the plate's small thickness.

The temperature was then converted to Young's modulus using an empirical relationship [203], which shows that Young's modulus decreases almost linearly within the investigated temperature range. The Poisson's ratio of the material may change slightly during heating [64], but its effect on the phase velocity of the A_0 mode is negligible, as shown earlier. Therefore, the usual value at room temperature was used for all temperatures. The material properties of a steel plate determined at room temperature are listed in Table 8.2.

Ultrasonic signals were measured on a 300 mm diameter circle around the heated area, with 100 array positions evenly spaced along the circumference. A PZT transducer was used to excite the A_0 mode by applying out-of-plane displacement at one position, while displacement was measured at other positions using a Polytec OFV-505 laser vibrometer. In each measurement, a 5-cycle Hanning windowed tone-burst signal at a center frequency of 60 kHz was generated by a Tiepie Handyscope HS3. Only transmitted signals were used, as reflections from smoothly varying defects are known to be negligible, and most of the scattering information remains in the transmission part of the signal [90]. Thus, the measurements were taken on half of the circle, with 51 measured points.

A contour plot of the typical measured out-of-plane displacement signals is shown in Figure 8.7(a). A gating function, as shown in Figure 8.7(b), was applied to remove unwanted components and obtain the first arrival wavepacket. This gating function is a rectangular window smoothed with rising and falling sinusoids on both sides. This process was repeated 100 times to build a matrix of 100×51 signals.

Similarly to the simulations, the measured signals in the experiment were calibrated before being used for inversion. However, because of the presence of the stiffness defect, signals from a homogeneous medium could only be obtained from limited positions. For calibration, the waveform was obtained from the position unaffected by the scatterer and then mapped to all other receivers according to the

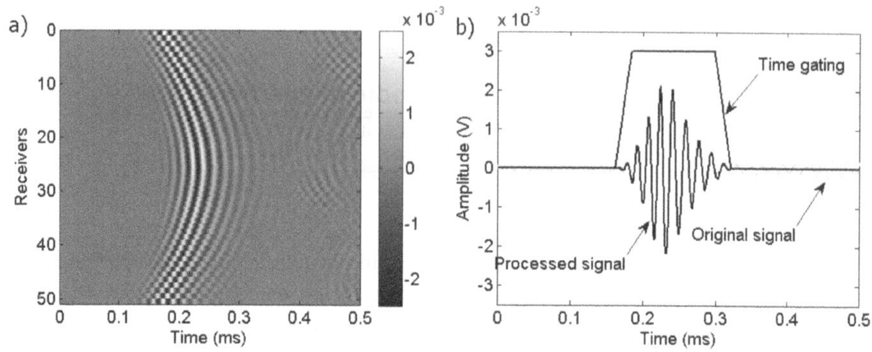

Figure 8.7 (a) Contour plot of the measured out-of-plane displacement of 51 measured signals; and (b) experimental time trace from the 26th receiver and the signal after time gating [187].

cylindrical spreading behavior of the wave propagation.

8.1.3.2 Validation

The temperatures of the heated plate were measured on a grid with dimensions of 400 mm × 300 mm, sampled with a spacing of 1.25 mm. The temperatures outside of this area were confirmed to remain constant and were the same as the temperatures measured on the perimeter of the grid. These temperatures were then converted to the corresponding Young's modulus values.

Young's modulus from the ultrasound measurements was derived from the phase velocity of the A_0 mode and reconstructed using the FWI. The calculation domain for the reconstruction was set to 400 mm × 400 mm, with a grid spacing of 1 mm. A sequential frequency group with frequencies of 45 and 60 kHz was used, and the calculation was initiated from a homogeneous background model. Up to 30 iterations were used to reach convergence.

The normalized Young's modulus reconstructed from the temperature measurements and the ultrasound measurements is depicted in Figures 8.8(a) and (b), respectively. The effect of temperature on Young's modulus is visible in both figures. Young's modulus rapidly decreases in the heated area, slightly offset from the center of the array, with a slightly asymmetric variation regarding its minima. The ultrasound reconstruction exhibits more noise compared to the temperature measurement. The region affected by temperature, where Young's modulus is significantly reduced, appears slightly larger and exhibits a slightly irregular shape in the ultrasound reconstruction. However, the region with the lowest values of the Young's modulus appears to be fairly similar in both reconstructions.

The general variation of the modulus in the homogeneous background in the ultrasound reconstruction remains below 5% of its mean value, with slightly greater

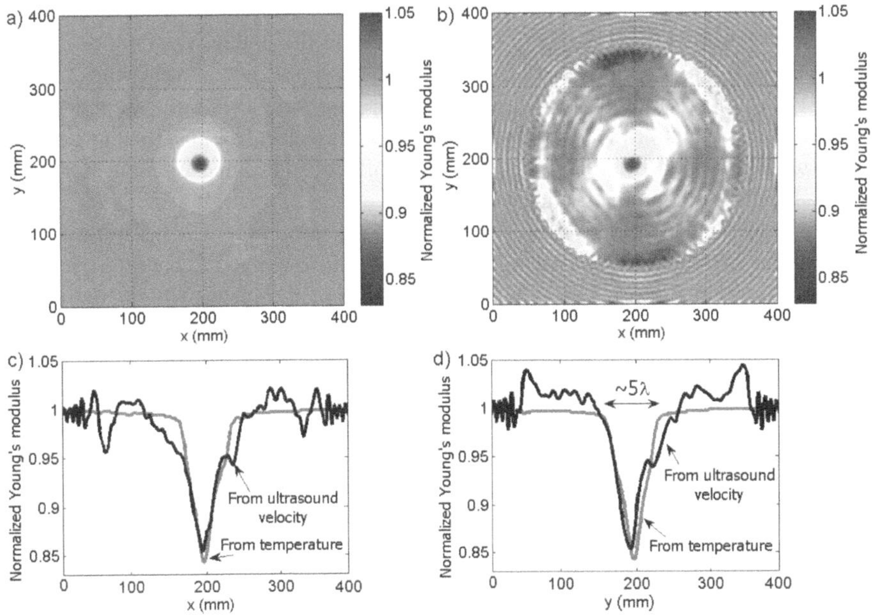

Figure 8.8 (Color online) Normalized Young's modulus: (a) Reconstructed from temperature; (b) reconstructed from ultrasound velocity; (c) horizontal cross-section; and (d) vertical cross-section [187].

variability observed at the positions of the array sensors. Several factors may contribute to this variability. One reason could be the velocity variation of the propagating guided wave in the plate. Measurement of the arrival times of the monitored signals, selected outside the influence of the heated area, showed that the group velocity varied to 4%. Similar arrival times were also measured using the same generator/receiver pairs in opposite directions, suggesting slight plate anisotropy. This leads to an underestimation or an overestimation of the reconstructed velocity in affected areas. Second, it was observed that the majority of the excitations did not produce ideal point source wave generation, resulting in moderate amplitude drifts from the desired waveform along the receivers. This introduces errors in the calibration process, leading to artifacts in the reconstruction. An additional factor could be the increase in temperature in the contact area between the transducer and the plate, which might affect the coupling. Although excitation was performed away from the high-temperature heating zone, the lower conductance of the coupling material can accumulate heat and may cause some directivity to excitation.

Figures 8.8(c) and (d) display the comparison between the reconstructions of Young's modulus along the cross sections with the largest drop in Young's modulus along the horizontal and vertical axes, respectively. In both figures, the temperature measurement reconstruction exhibits a regular-shaped Young's modulus curve, whereas the ultrasound reconstruction appears to be more irregular. However, it can be clearly observed that below the value of 0.95 (outside the noisiest region) of the normalized Young modulus, there is a very good agreement in capturing the smallest value of Young's modulus between the two approaches. This agreement was expected, as the extent of the Young's modulus reduction area (5λ) is much larger than the size from which the reconstruction error of ultrasonic measurement begins to grow (1.5λ), as shown in simulations.

The slight underestimation of Young's modulus drop obtained from ultrasound data can be attributed to the heating effect on the thickness and density of the plate. The rise in temperature causes thermal expansion of the plate, slightly increasing the thickness and reducing the density, which has not been considered in the model. Both changes result in an increase in velocity, which is the opposite effect due to the reduction of Young's modulus. This velocity change for the steel plate at the highest temperature was quantified using a dispersion equation and was found to be around 15% from the change due to Young's modulus reduction. This indicates that the Young's modulus is still the dominant factor for the velocity change and demonstrates the applicability of the method to determine the variation of Young's modulus.

8.2 ANALYSIS OF ANISOTROPIC EFFECTS

8.2.1 BACKGROUND

UGWT is a powerful method for assessing material properties by analyzing ultrasonic signals captured by a transducer array around an inspection area. It offers a quick and efficient way to assess large, challenging-to-reach areas, making it valuable for various applications. However, most UGWT research has focused mainly on isotropic materials, such as metals. Many high-performance applications involve anisotropic materials and structures, such as composite pressure vessels [28], pipe elbows [25], and anisotropic stainless steel pipes [206]. The application of UGWT to anisotropic waveguides comes with its own set of challenges, including complex wave behavior due to directional dependence and intricate defect geometries [115,167].

To evaluate anisotropic waveguides using UGWT, it is crucial to accurately model the velocity variations induced by anisotropy [63], which plays a vital role in obtaining precise imaging results. The reconstruction of defects in UGWT is based on solving an inverse problem, which is based on a suitable forward model. The choice of an appropriate forward model should accurately describe wave propagation behavior in the structure and be sensitive to property changes.

Although 2D acoustic models have been widely used in UGWT due to their balance between computational complexity and cost, most previous studies focused on isotropic materials. Anisotropic materials introduce greater complexity, as wave propagation and velocity depend on the stiffness of the material, characterized by numerous constants [255].

Recent efforts have introduced models such as the elliptically anisotropic acoustic forward model, designed for UGWT in pipe bends [23,24]. However, these models may not capture the complexity of wavefronts in cases such as guided waves in cross-ply multilayered composite plates.

This study aims to investigate the applicability of the 2D acoustic model in UGWT for anisotropic plates. The model incorporates approximate Thomsen-style anisotropic parameters to represent artificially inhomogeneous and anisotropic velocity fields [237, 241]. It is used in a Full Waveform Inversion (FWI) algorithm to map velocity variations resulting from modeled stiffness defects in multilayered composite plates with mild to moderate anisotropy. The focus is on the low-frequency anti-symmetric Lamb wave mode A_0 due to its sensitivity to stiffness variations.

The chapter proceeds to describe the methods for acoustic forward modeling, its implementation for the guided wave problem, and FWI in Section 8.2.2. It provides insights from numerical studies, including Finite Element (FE) and acoustic modeling, and presents the results of forward modeling and defect imaging in Section 8.2.3. The subsequent section, Section 8.2.4, covers an experimental study, including measurement procedures and a validation experiment. The chapter concludes with a discussion of the effects of anisotropy in the plate and defects on the accuracy of inversion and the limitations in UGWT in Section 8.2.5.

8.2.2 METHODS

8.2.2.1 Acoustic forward modeling

Consider a composite plate, as shown in Figure 8.9. We assume that the behavior of a guided wave propagating through a plate with varying stiffness is akin to an acoustic wave traveling in a 2D medium with variable velocity. The acoustic wave is modeled in transversely isotropic media with a tilted symmetry axis. The propagation of the wave of this medium is described by a coupled system of second-order partial differential equations [276]:

$$\begin{cases} \frac{1}{v_0^2}\frac{\partial^2 p}{\partial t^2} - (1+2\delta)Hp - H_0 p = (1+2\delta)Hq \\ \frac{1}{v_0^2}\frac{\partial^2 q}{\partial t^2} - 2(\varepsilon-\delta)Hq = 2(\varepsilon-\delta)Hp. \end{cases} \tag{8.4}$$

In these equations, p represents the pressure wavefield, q is an auxiliary wavefield, and v_0 is the velocity of the pressure wave along the symmetry axis. The parameters ε and δ are anisotropic parameters defined by Thomsen [237], and θ is the angle of the symmetry axis relative to the x-axis. When $\varepsilon = \delta$, the model becomes elliptically anisotropic. For $\varepsilon = \delta = \theta = 0$, the equations reduce to the second-order acoustic isotropic wave equation. The cross-derivative terms in the differential operators H and H_0 account for the rotation of the symmetry axis and are defined as:

$$\begin{cases} H = \cos^2\theta\frac{\partial^2}{\partial x^2} + \sin^2\theta\frac{\partial^2}{\partial y^2} - \sin 2\theta\frac{\partial^2}{\partial x \partial y} \\ H_0 = \sin^2\theta\frac{\partial^2}{\partial x^2} + \cos^2\theta\frac{\partial^2}{\partial y^2} + \sin 2\theta\frac{\partial^2}{\partial x \partial y}. \end{cases} \tag{8.5}$$

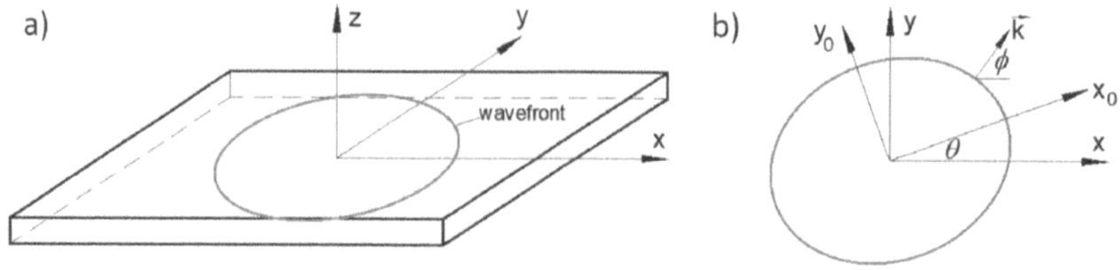

Figure 8.9 (a) Acoustic model of guided wave propagation in a composite plate; (b) Top view of the wave-field with symmetry axis (x_0, y_0), tilt angle θ, wavenumber k, and phase angle ϕ [188].

It is important to note that these simplified equations, neglecting shear waves, still accurately describe travel time, geometrical amplitude, and wavefront spreading. The kinematic behavior of propagating waves with a known symmetry tilt angle can be characterized by three model parameters: v_0, ε, and δ. It is worth mentioning that the solution of Equation (8.4) becomes unstable when $\varepsilon < \delta$ due to unstable shear waves. To mitigate this instability, frequency-domain modeling with absorbing boundary conditions can be employed [152].

8.2.2.2 Anisotropic parameters

In acoustic media, velocity and anisotropic parameters can be directly determined from the stiffness constants of the medium. However, this direct relation does not exist in the case of a plate. In situations with weak anisotropy, we can match the acoustic velocity model to the guided wave model by using approximated Thomsen parameters. These parameters can be estimated by considering the first-order terms of the linearized phase velocity v_{ϕ_0} at a few discrete phase angles as follows:

$$\begin{aligned} \varepsilon &= \frac{v_{90}-v_0}{v_0} \\ \delta &= 4[\frac{v_{45}}{v_0} - 1] - \varepsilon. \end{aligned} \tag{8.6}$$

Here, v_0 represents the phase velocity along the symmetry axis at $\phi_0 = 0$, while v_{45} and v_{90} are the phase velocities at the phase angles $\phi_0 = 45^\circ$ and 90°, respectively. The parameter ε describes the relative velocity difference along the symmetry axes of the problem, while δ characterizes the velocity variation at intermediate phase angles.

8.2.3 NUMERICAL STUDY

8.2.3.1 Finite element modeling

Wave propagation simulations were conducted in Abaqus Explicit, exploring two configurations. In the first, waves were excited at the center of the composite plate, monitored by a circular array (Figure 8.10(a)). The second configuration involved

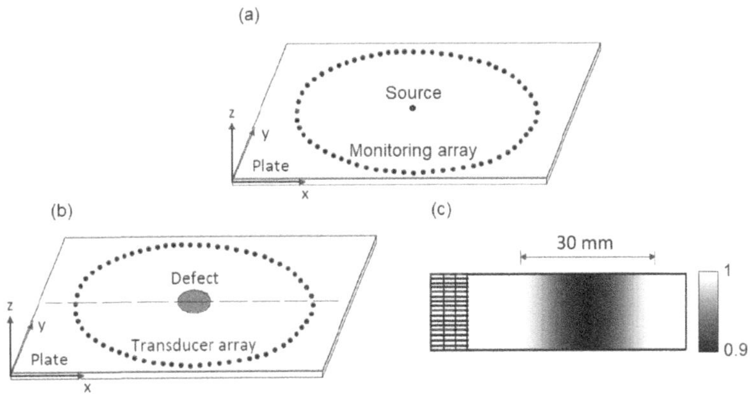

Figure 8.10 (a) Plate setup with a central source and a circular monitoring array surrounding it; (b) Plate setup with a central stiffness defect and a circular transducer array surrounding it; (c) Cross-section of the meshed plate with a through-thickness defect [188].

Table 8.3
Stiffness constants (in GPa) of the SE84LV prepreg ply used in the present study; the mass density is 1540 kg/m^3; the ply thickness is 0.15 mm [188]

C_{11}	C_{12}	C_{13}	C_{22}	C_{23}	C_{33}	C_{44}	C_{55}	C_{66}
107.8	4.51	4.51	10.31	4.24	10.31	3.04	4.45	4.45

transducers around a circular stiffness defect (Figure 8.10(b)). In both cases a 200 mm diameter array with 40 transducers was used.

The composite plates, comprising 16 plies of carbon fiber/epoxy (Table 8.3), were modeled in FE modeling. Each ply was individually modeled and tied together at interfaces. The plates were meshed using 8-node brick elements (C3D8R). Three stacking sequences ($[0/45/90/\text{-}45]_{2S}$, $[0/90]_8$, $[0]_{16}$) were considered to study anisotropy. Absorbing layers minimized the reflections from the edges of the plate.

Common types of damage, such as local stiffness reduction, were investigated using artificial defects. The defect (Figure 8.10(c)) was an axisymmetric defect modeled with reduced stiffness or with a change in the anisotropy of the material. In the stiffness defect, the constants were uniformly reduced with a Hann-shaped variation. Anisotropic defects involve the smooth change in fiber orientation through the thickness. The study aimed to understand the effects of simple artificial defects.

A 5-cycle Hann-windowed tone-burst signal at 50 kHz simulated the input signal. The A_0 mode was generated at a source point, measuring the wavefields at other receivers. This process created a 40×39 signal matrix, transformed into the frequency domain for inversion with extracted and calibrated components.

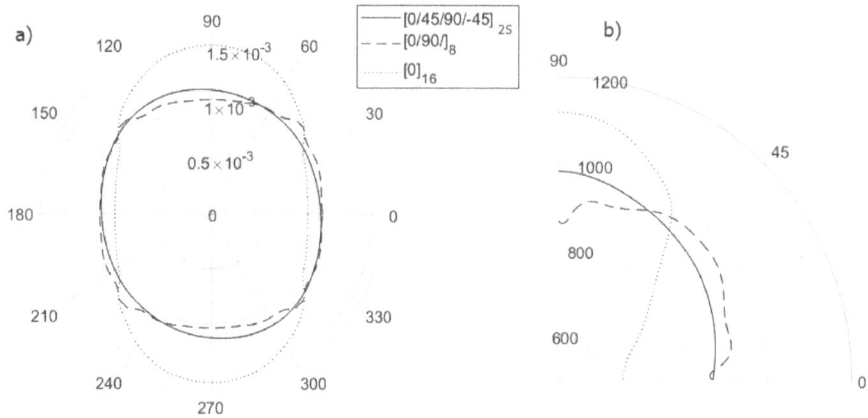

Figure 8.11 Characteristic wave curves for A_0 mode at 50 kHz for different composite lay-ups: (a) slowness $s(\phi) = k(\phi)/\omega$; (b) phase velocity $c_{ph}(\phi)$ [188].

8.2.3.2 Acoustic modeling

Acoustic modeling of plate-like guided wave propagation relies on prior knowledge of the behavior of the guided wave mode. The slowness curves for the A_0 mode at 50 kHz (Figure 8.11(a)) were obtained from FE simulations. Anisotropic wave propagation is evident for all composite lay-up cases. The nearly quasi-isotropic plate ($[0/45/90/\text{-}45]_{2S}$) exhibits a slightly elliptical and asymmetric slowness profile. The cross-ply plate ($[0/90]_8$) shows a rectangular shape with the highest slowness values at 45° from the symmetry axes. The unidirectional composite plate ($[0]_{16}$) shows a highly elongated slowness along 90°.

The velocity curves (Figure 8.11 (b)) in the first quarter of the coordinate system highlight the variable phase velocities for all plates. Acoustic modeling parameters (tilt angle θ, phase velocity along symmetry axes x_0, and Thomsen parameters ε and δ) are presented in Table 8.4. An alternative choice of main-symmetry axis can yield negative Thomsen parameters.

Acoustic simulations in the FD method employed a 250 × 250 grid with a 1 mm spacing. The starting models for inversion were homogeneous backgrounds of v_0, ε, and δ. For the unidirectional composite plate (where $\varepsilon > \delta$), a small, smoothly tapered circular region with $\delta = \varepsilon$ was established at the source point to eliminate unwanted shear wave excitation [60]. This introduced minimal kinematic error in non-horizontal or non-vertical directions.

8.2.3.3 Forward simulation results

In the simulation, wave propagation was analyzed in three composite plates (Table 8.3) with a source at the center of the circular matrix (diameter: 200 mm). Figure 8.12 compares snapshots in the time domain of propagation of the A_0 mode at 50 kHz

Table 8.4
Wave simulation parameters in acoustic modeling [188]

Composite lay-up	θ (°)	v_0 (m/s)	ε	δ
$[0/45/90/\text{-}45]_{2S}$	108	868	0.13	0.13
$[0/90]_8$	45	866	0	0.45
$[0]_{16}$	90	654	0.71	0.35

from FE and FD simulations. The horizontal axis aligns with the x-axis, the radial coordinate represents the propagation time, and the contour color indicates the plate displacement amplitude. Good agreement is observed in wave fronts for all plate cases, irrespective of the propagation direction. However, amplitude mismatches are observed.

For the cross-ply laminate, FE shows wave energy along the principal axes, while FD displays nonsymmetric energy distribution focused at 135° and 315°. In the unidirectional composite plate, FE concentrates wave energy along the fiber direction, while FD shows a more equal distribution in all directions. This suggests reduced accuracy in forward modeling with increasing medium anisotropy. Overall, this numerical example affirms the ability of the 2D acoustic model to predict guided

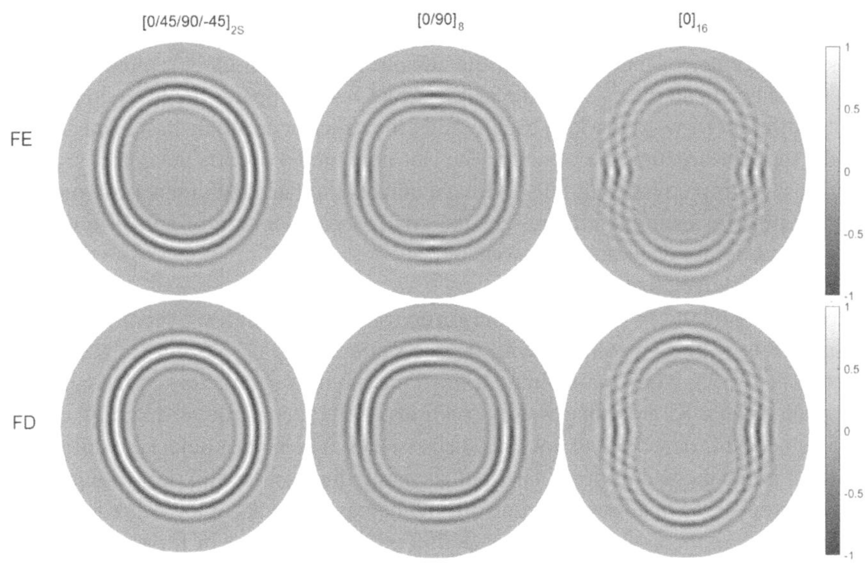

Figure 8.12 Forward modeling results of the A_0 mode excited at 50 kHz obtained from FE and FD simulations for different composite lay-ups [188].

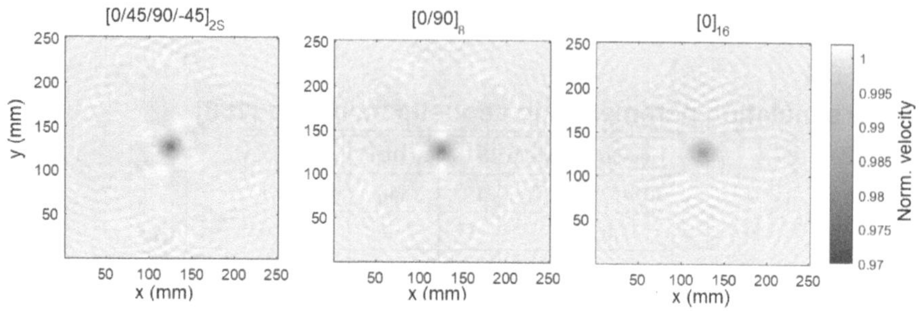

Figure 8.13 FWI results with an anisotropic forward model using A_0 mode at 50 kHz: normalized velocity v_0 maps for different composite lay-ups with central 30 mm Hann-shaped stiffness defect with maximum loss in stiffness constants of 10% [188].

waveforms in composite plates with weak to moderate anisotropy.

8.2.3.4 Inversion results

In this case, we assess the reconstruction of the defect in composite plates using either isotropic or anisotropic forward models in FWI. Reconstructed velocity maps, obtained by inverting FE data at 50 kHz, start from a homogeneous background velocity along the symmetry axis and Thomsen parameter models, using 40 iterations for inversion. The isotropic model sets the Thomsen parameters of the background to zero. Figure 8.13 displays the well-reconstructed 30 mm wide defect with stiffness loss in the center of the plate using the anisotropic forward model, maintaining its circular shape in all the cases of the plate. Figure 8.14 compares the inversion results using different forward models, highlighting the superior accuracy of the anisotropic model in reconstructing the velocity. The isotropic model distorts the defect shape in the quasi-isotropic plate and fails to image defects in plates with increasing material anisotropy. This example emphasizes the crucial consideration of anisotropy in the forward model for accurate inversion.

Effect of defect size

Next, employing FWI with the anisotropic forward model, we reconstruct stiffness defects with diameters varying from 15 mm ($\simeq 0.9\lambda$), 30 mm ($\simeq 1.8\lambda$), and 60 mm ($\simeq 3.5\lambda$), while maintaining a fixed maximum stiffness loss in the center of the plate. Figure 8.15 compares cross sections of the reconstructed velocity and the true velocity for defects with varying surface sizes. The largest defect is well reconstructed in the quasi-isotropic plate, but accuracy diminishes with decreasing defect size. Moreover, as plate material anisotropy increases, the accuracy degrades, aligning with the findings of the previous section on the challenges of forward modeling more anisotropic materials. Although tomographic imaging ideally reconstructs velocity variations down to $\lambda/2$ with a perfectly matching forward model, the acoustic forward model, which does not encompass all wave physics of guided wave propagation, leads to poorer resolution and accuracy. This aligns with similar findings in

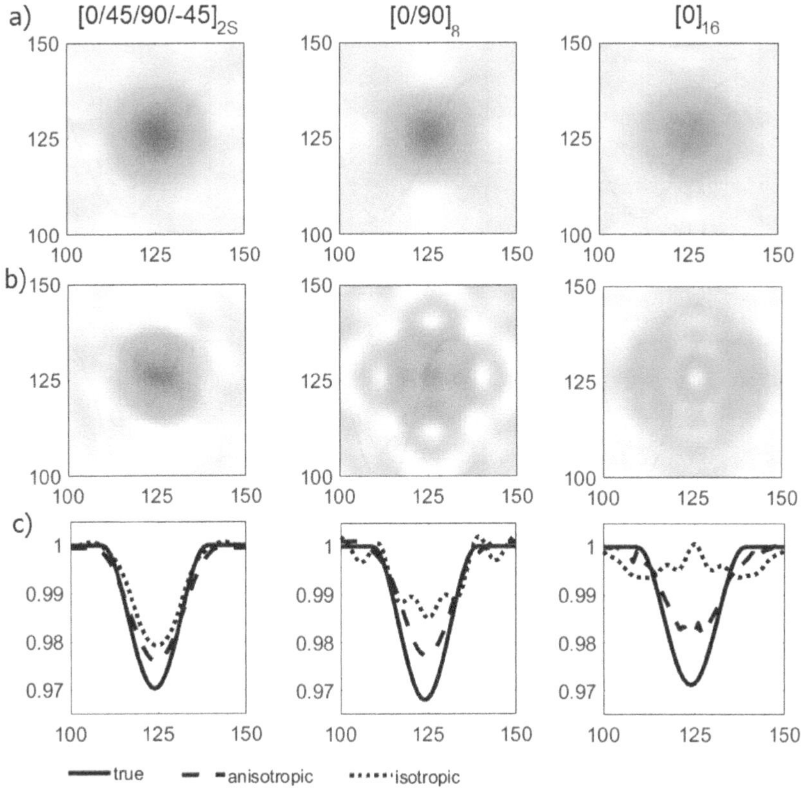

Figure 8.14 Comparison of the reconstructed velocity maps using different forward modeling: (a) anisotropic forward model, (b) isotropic forward model, (c) cross-section of the defect at y = 125 mm [188].

studies with isotropic materials, where velocity variations around 2λ can be reconstructed using acoustic forward models [172].

Anisotropic effect

In the final case, we explore the velocity reconstruction of anisotropic defects, such as delamination in a composite laminate. Unlike previous cases where only the stiffness of the defect was slightly reduced without significant changes in Thomsen parameters, here we investigate defects where Thomsen parameters are considerably altered. The defects, 30 mm in width and offset by 40 mm from the center in both coordinate directions, include one in the quasi-isotropic plate and another in the unidirectional plate. The orientation of the fiber direction inside the defect smoothly changes to match the surrounding plate's fiber direction, resembling a unidirectional lay-up as $[18]_{16}$ in the quasi-isotropic plate. The second defect in the unidirectional plate has a lay-up in the center as $[-18/27/72/-63]_{2S}$. Both defects exhibit nontrivial rotation angles to align with the symmetry axes of the defect and the surrounding

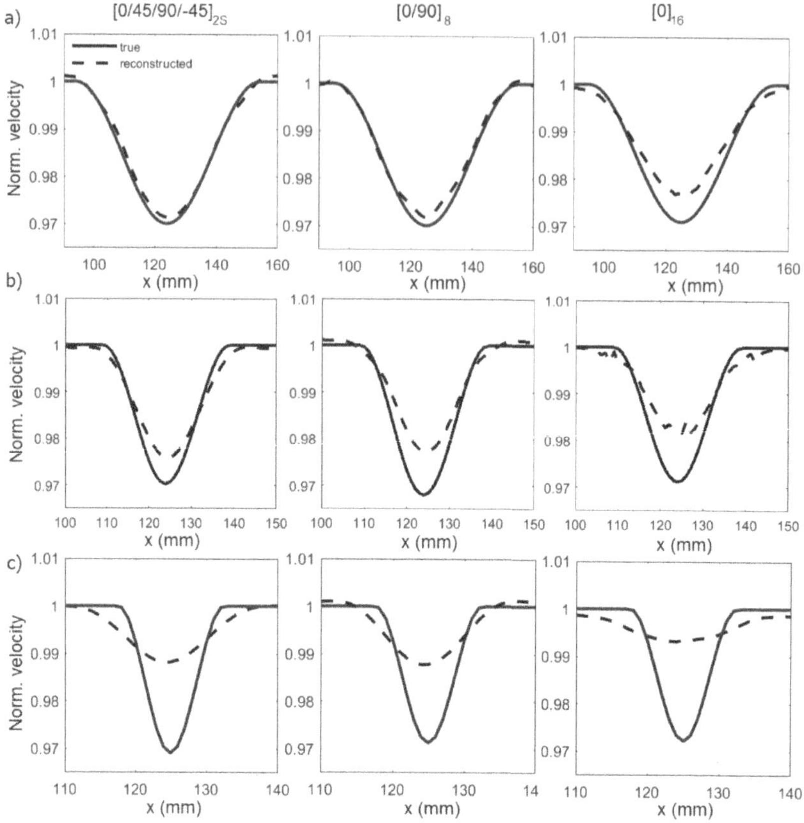

Figure 8.15 FWI results with A_0 mode at 50 kHz: cross-section of the normalized velocity v_0 profile of the defect at the plate center for different composite lay-ups with varying surface diameter of (a) 60 mm, (b) 30 mm, and (c) 15 mm [188].

plate in acoustic modeling. The results in Figure 8.16(a) show detectable defects, but their shapes become more irregular and artifacts appear in the surrounding image. Figure 8.16(b) compares reconstructed velocity with the actual velocity, revealing a decrease in accuracy compared to previous cases. This challenge arises from the strong anisotropy, causing a loss in accuracy due to the increased ill-posedness in the inversion. The coupling between parameters intensifies the sensitivity of the inversion, leading to significant changes from one parameter to the next [249].

8.2.4 EXPERIMENTAL STUDY

8.2.4.1 Specimen and measurements

The experimental setup involved a composite plate measuring 500 mm × 500 mm × 2.4 mm with a layup of $[0/45/90/\text{-}45]_{2S}$ and properties that matched the simulation

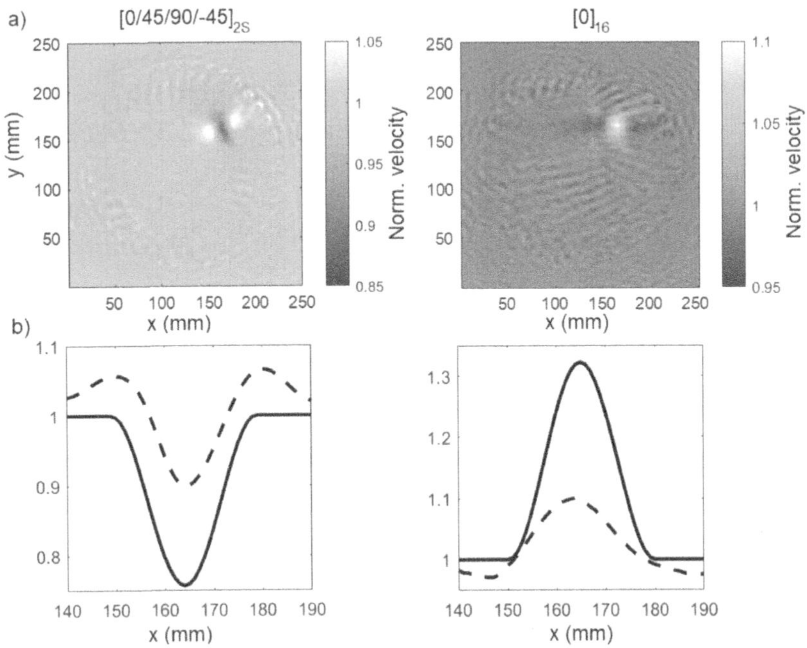

Figure 8.16 FWI results with A_0 mode at 50 kHz: (a) normalized velocity v_0 maps for different composite lay-ups with noncentral 30 mm Hann-shaped anisotropic stiffness defect; (b) cross-section of the defect at y = 165 mm [188].

conditions (Table 8.3). Three separate ply laminates with layups [0/45/90/-45], [0/45/90/-45]$_S$, and [-45/90/45/0] were initially prepared and cured in an autoclave. A 30 mm diameter hole was cut into the center of the second laminate, filled with resin Epolam 2051 + 3M glass micro balloons. The laminates were bonded with the adhesive film, forming a plate with a reduced area of stiffness defect in the center (Figure 8.17(a)).

The experimental setup (Figure 8.17 (b)) included a 40 element matrix (200 mm diameter) for ultrasound tomography measurements. A Doppler PZT transducer with a center frequency of 70 kHz, coupled to the composite plate, excited the A0 mode, while a Polytec OFV-505 laser vibrometer measured the surface displacement. Each measurement used a 5-cycle Hann-windowed tone-burst signal (centered at 60 kHz) generated by a Tiepie Handyscope HS3. Measurements were taken at 21 points in the transmission zone, covering half of the circle. A contour plot of typical out-of-plane displacement signals is shown in Figure 8.17(c), and a gating function (Figure 8.17(d)) was applied to isolate the first arrival wavepacket. This process was repeated for 40 excitation points, resulting in a signal matrix of 40 × 21.

8.2.4.2 Validation

The reconstructed velocity maps were generated by inverting the data at 50 kHz. The process involved starting from homogeneous background velocity along the

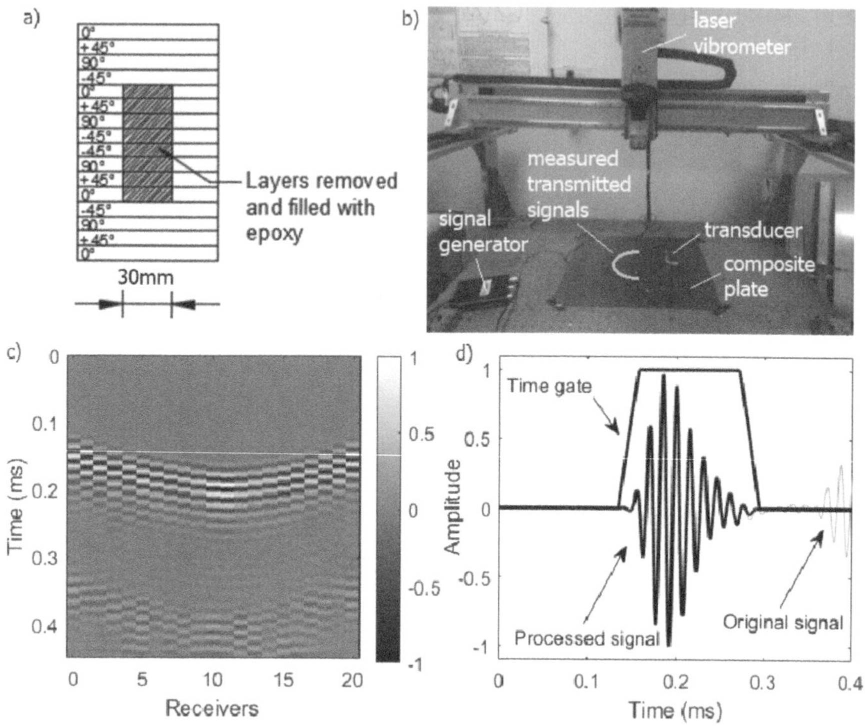

Figure 8.17 Experimental measurements: (a) cross-section of the composite plate with a stiffness defect; (b) setup to measure ultrasound propagation in a composite plate; (c) contour plot of the measured out-of-plane displacement of 21 measured signals; (d) experimental time trace from the 7th receiver and the signal after time gating [188].

symmetry axis and Thomsen parameter models, with 40 iterations for the inversion. Figures 8.18(a) and (b) present the normalized velocity reconstructed from experimental measurements and FE modeling, respectively. The defect area in the center of the plate exhibits a noticeable drop in velocity. Although the defect appears circular in the experiment, the FE modeling depicts a slightly elliptical shape elongated along the y-axis. Cross-sectional comparisons between the reconstructed velocity and the true velocity are illustrated in Figures 8.18(c) and (d). The velocity drop is slightly overestimated in the experiment, whereas it is accurately reconstructed from the FE data. The overestimation is attributed to the lower stiffness and reduced thickness of the plate in the defect area, possibly caused by curing shrinkage. In addition, this defect is more complex than the one used in FE, featuring smoothly changing material properties. It divides the waveguide into three parts, with a central layer exhibiting significantly lower acoustic impedance, leading to energy trapping and additional scattered wave fields not encountered in the acoustic model [145].

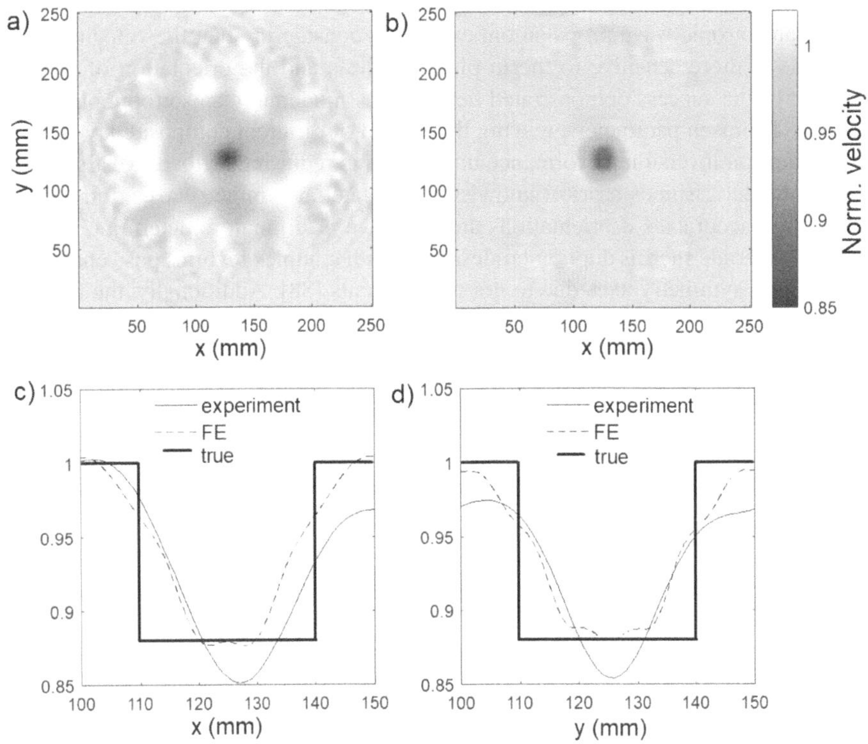

Figure 8.18 Normalized velocity obtained with FWI using the A_0 mode at 50 kHz: (a) experiment, (b) FE, (c) horizontal cross-section, and (d) vertical cross-section [188].

8.2.5 DISCUSSION

The study's results showcase the efficacy of a straightforward acoustic forward model in producing high-resolution velocity images in anisotropic plate-like waveguides through FWI. Thomsen parameters serve as a valuable link between true guided-wave velocities and those of the acoustic model, which is particularly beneficial in applications where Thomsen parameters are readily determined for defect characterization. However, as demonstrated, the accuracy of the reconstructions decreases with increasing background material anisotropy and the complexity of the defect.

One reason for the decreased accuracy is the diminishing accuracy of amplitude modeling in forward modeling, despite the accurate phase prediction. Mitigation strategies involve considering only phase information in the misfit minimization criteria, applying heuristic amplitude corrections [249], or modifying the directivity weights of the pressure source for the correct waveform amplitudes [9]. It is crucial to acknowledge that this reconstruction error cannot be entirely eliminated because of the acoustic assumption and the model's focus on weakly anisotropic media.

Several limitations merit consideration. The method may be less suitable for highly anisotropic wave propagation, especially considering that the velocity of the S_0 mode is more sensitive to the in-plane stiffness and the orientation of the layers [155]. The success demonstrated here is for a moderately anisotropic plate with specific Thomsen parameters, leaving the impact of different composite lay-up configurations on inversion performance unknown and requiring further investigation.

The model assumes a priori knowledge of material symmetry axes, introducing potential inaccuracies if orientations are unknown or difficult to determine. For example, materials such as duplex stainless steel with complex textures pose challenges in selecting symmetry axes due to grain alignments [88]. Additionally, the assumption that wave propagation has two symmetry planes may not hold for certain scenarios, such as the flexural mode in unbalanced laminates. Evaluating the symmetry of the flexural stiffness can provide insight into the validity of this assumption, considering that the wave velocity is proportional to the stiffness [129].

8.3 GUIDED WAVE TOMOGRAPHY OF PIPE BENDS

8.3.1 INTRODUCTION

Pipe bends in pipeline systems face unique challenges due to stress from changes in fluid direction, making them vulnerable to fatigue, corrosion, and cracking. Fatigue causes cracks from repeated stress cycles, while corrosion thins pipe walls, weakening their ability to withstand pressure. Cracks, which can result from both fatigue and corrosion, are especially common in high-stress areas, such as bends. These cracks, if not detected, can compromise the integrity of the pipe and cause failure. Monitoring damage in pipe bends is critical to maintaining pipeline efficiency, reducing maintenance costs, and preventing catastrophic failures.

Traditional NDT methods, such as X-rays, thickness gauges, and in-pipe robots, have limitations. X-rays can be cumbersome, thickness gauges are slow for large areas, and in-pipe robots require halting operations, leading to downtime.

UGWT offers a promising alternative for damage detection. UGWT uses transducer arrays to generate and receive guided waves, allowing for the creation of a thickness map of the pipe. By analyzing the waveforms, changes in wall thickness can be detected, with variations in phase velocity corresponding to thickness loss.

GWT has been successfully applied to simple structures like plates and straight pipes. For straight pipes, guided wave propagation can be modeled as 2-D, with an acoustic wave equation similar to that used for plates. However, the wavefield in pipe bends is more complex as a result of the anisotropic effects of the bend radius, making the modeling of wave propagation more challenging.

Previous studies on pipe bends have focused on 2-D models. Volker and van Zon [251] used a 2-D deformed grid representation to reconstruct defects numerically, while Brath et al. [23] applied an orthogonal 2-D rectangular domain to simulate wave propagation. This model created an artificially inhomogeneous and elliptically anisotropic wavefield. These approaches, along with travel-time tomography, have been used to reconstruct thickness maps in bends affected by corrosion. Travel-time tomography, while effective, has limitations, including its reliance on time-of-flight

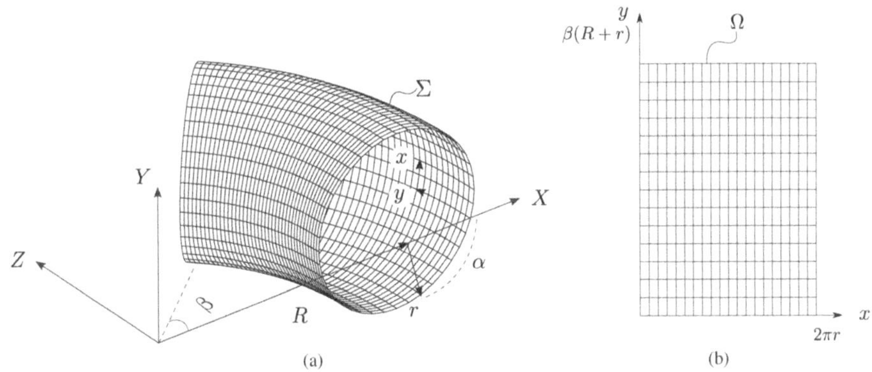

Figure 8.19 A torus section, representing the bend of the pipe, is tranlsated from (a), the 3-D space domain Σ to (b), the 2-D space domain Ω [184].

measurements and its inability to capture diffraction effects. In pipe bends, the focusing effect can cause rays to bypass the bend elbow, reducing the accuracy of the tomography, especially depending on the circumferential position of the defects. To address this, Brath et al. [25] added extra transducers along the bend elbow to improve coverage and reduce dependence on the position of the defect.

In this chapter, we introduce a new 2-D acoustic model for pipe bends which is based on the finite difference method and incorporates the anisotropy of the bend's curvature [184]. The method is validated experimentally and numerically.

8.3.2 2-D ACOUSTIC FORWARD MODEL OF A PIPE BEND

8.3.2.1 Orthogonal Parametrization

Consider the section of the torus from Figure 8.19 (a), defined in the 3-D space domain Σ, with mid-thickness or central radius r_c, bend radius R, azimuth longitude of the torus β around the Y-axis, and torus latitude α respecting the plane XY. To model the anisotropic wave propagation around the bend in the 2-D space domain Ω shown in Figure 8.19 (b), a translation of the torus section from Σ to Ω is needed. We followed the parameterization proposed by Brath et al. [23]. According to this, the torus section is unwrapped from the longest radius path of the bend (in extrados position $R+r_c$). Second, the 2-D horizontal and vertical axes x and y were set equal to the torus circumference $2\pi r_c$ and the extrados azimuth length of the bend $\beta(R+r_c)$, respectively. As a result, the 2-D domain Ω can be expressed as $\Omega = [0, 2\pi r_c] \times [0, \beta(R+r_c)]$, with the extrados position located in the middle of the circumference at $(\pi r_c, y)$. The intrados position is located at the beginning of the circumference at $(0, y)$ and at the end $(2\pi r_c, y)$.

The orthogonal parameterization of the torus is given by the set of equations as

$$\begin{aligned} X &= (R + r_c \cos \tfrac{x}{r_c}) \cos \tfrac{y}{R+r_c} \\ Y &= r_c \sin \tfrac{x}{r_c} \\ Z &= (R + r_c \cos \tfrac{x}{r_c}) \sin \tfrac{y}{R+r_c}, \end{aligned} \tag{8.7}$$

where two-dimensional coordinates $\{x,y\}$ are used to express three-dimensional space coordinates $\{X,Y,Z\}$. In this way, 3-D space coordinates can be mapped directly into the 2-D domain, and vice versa.

8.3.2.2 Wave Equations

Given the orthogonal parameterization of the pipe bend in Equation 8.7, the propagation of guided waves in the 2-D domain can be modeled as an elliptically anisotropic wavefield [23]. Consequently, the acoustic wavefield of a torus section can be represented as a vertical transversely isotropic (VTI) medium. The acoustic wavefield of a VTI medium can be described by a system of second-order partial differential equations (8.4) and (8.5).

To describe the elliptical anisotropy wavefield for a torus section as a VTI medium, consider the relationship between the phase velocity and the velocity along the symmetry x-axis, denoted as v_v [276]:

$$\frac{cos^2\overline{\theta}}{v_v^2} = \frac{1}{v^2} - (1+2\delta)\frac{\frac{sin^2\overline{\theta}}{v_v^2}\frac{1}{v^2}}{\frac{1}{v^2} - 2(\varepsilon-\delta)\frac{sin^2\overline{\theta}}{v_2}}, \tag{8.8}$$

where $\overline{\theta}$ is the angle with respect to the symmetry axis. Notably, for a VTI medium, $\overline{\theta} = \pi/2$, and given the elliptical anisotropy condition, $\varepsilon = \delta$ [237]. In addition, following the orthogonal representation of the bend from Equation 8.7, the velocity along the symmetry x-axis v_v is determined by $v_v = \frac{1}{\rho_e}$, with $\rho_e = \frac{R+rcos(\theta)}{R+r_c}$ [23]. Hence, Equation 8.8 yields $\varepsilon = \delta = \frac{1}{2}(\frac{1}{\rho_e^2} - 1)$.

8.3.3 NUMERICAL MODELING

8.3.3.1 Configuration of the problem

In this work, we considered the steel tube bend geometry described in Figure 8.20. The geometry consists of an inner radius r_{in}= 0.1015 m, an outer radius r_{out}= 0.1095 m, an azimuth longitude $\beta = \pi/2$ with a major radius R = 0.329 m, and two straight pipe sections at the beginning and end of the bend that are each 1 m long. Straight sections were only included in the elastic modeling and experimental measurements to avoid reflections due to the edges. Twenty transducers were equally distributed between each transducer ring and located at both ends of the bend, as shown in Figure 8.20(c).

8.3.3.2 Acoustic modeling

The image reconstructions for numerical and experimental analysis are computed by considering all wave packets that have completed up to three full revolutions around the circumference. As a result, the 2-D domain consists of three identical replicas, as illustrated in Figure 8.21. Each replica is discretized using 4.143 × 4.143 mm^2 square elements, leading to 481 × 166 nodes (circumferential-azimuthal). Naturally,

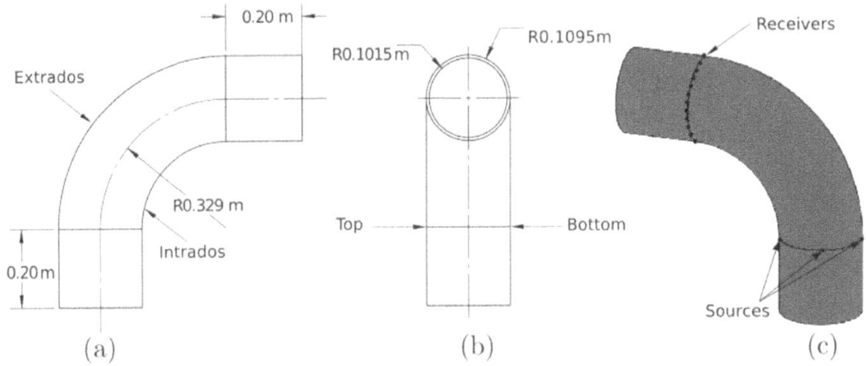

Figure 8.20 Schematic of the pipe bend. (a) top view, (b) lateral view, and (c) isometric view of the simulated pipe bend [184].

the emissions were located in the central replica, while the transmissions were also located in the additional replicas; this was done to include higher-order helical wave paths in the simulation [23].

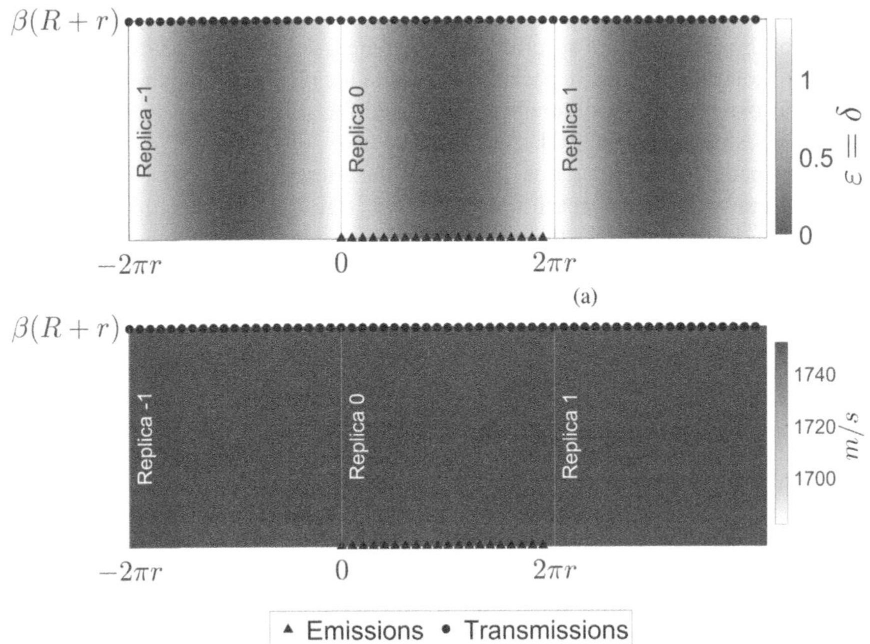

Figure 8.21 Input parameters for the acoustic modeling. Thomsen parameters ε and δ (a) make the wavefield artificially anisotropic given the phase velocity (b) along the bend [184].

Figure 8.21(a) and (b) show the distribution of Thomsen parameters along the bend and the phase velocity distribution or background velocity, respectively. The phase velocity remains constant as a result of the isotropic nature of the steel. However, the wavefield is artificially made anisotropic by the Thomsen parameters. The anisotropy extends from the center of the replica towards the edges, as shown in Figure 8.21(a).

We followed the FD method using the mixed grid approach for simulations in the acoustic domain [152]. We carried out calculations using the TOY2DAC 2-D frequency domain engine [136]. TOY2DAC uses the optimization routines from the SEISCOPE toolbox to address the minimization problem in the frequency domain for a given frequency. The fast Fourier transform was used to convert the results from the time domain to the frequency domain. The phase velocity was the inverted parameters at a specific frequency, and we derived the remaining thickness maps by interpolating the reconstructed velocity as a function of the frequency-thickness product using the dispersion curve of the A_0 mode. Further details on the 2-D acoustic forward model that uses Thomsen parameters to map the elliptical anisotropy of the bend area, as well as its similarity to its 3-D elastic modeling counterpart and scattering, can be found in [183].

8.3.3.3 Helical path separation

The 2-D acoustic model, defined in a plane, does not account for the cyclic characteristic of the pipe within a specific inspection area, defined by a pair of transducer rings. To isolate the first wave packet that reaches the receivers from the elastic modeling and experimental measurements, we used the robust helical path separation algorithm, as developed in [95].

In the helical path separation algorithm, the time traces $u_{r_v,s}(t)$, which originate from emission s and reach transmission r_v, are duplicated m times to create $u_{r_v+Nm,s}(t)$. Here, N denotes the total number of sources, m the number of replicas, and Nm the increment in each receiver transducer number. Next, the replicated time traces $u_{\tilde{r}_v,s}$ are time-reversed by

$$\overleftarrow{U}_{\tilde{r}_v,s}(\omega) = FFT(u_{\tilde{r}_v,s})exp[-ik(\omega)x_{\tilde{r}_v,s}], \tag{8.9}$$

where $\overleftarrow{U}_{\tilde{r}_v,s}(\omega)$ is the time-reversed signal in the frequency domain, $k(\omega)$ is the wave number for each frequency ω, and $x_{\tilde{r}_v,s}$ is the distance traveled from transmission to emission.

Moreover, a series of bandpass filters and windows are applied to extract the first wave packet arriving $\overleftarrow{\Phi}_{\tilde{r}_v,s}(\omega)$. Then, $\overleftarrow{\Phi}_{\tilde{r}_v,s}(\omega)$ propagates forward:

$$\overrightarrow{\Phi}_{\tilde{r}_v,s}(\omega) = \overleftarrow{\Phi}_{\tilde{r}_v,s}(\omega)exp[ik(\omega)x_{\tilde{r}_v,s}]. \tag{8.10}$$

Finally, the original replicated time traces $u_{\tilde{r}_v,s}$ are cleaned up by minimizing the waveforms belonging to other replicas. Mathematically, this is represented as

$$u^{hp}_{\tilde{r}_v,s}(t) = u_{\tilde{r}_v,s}(t) - \overrightarrow{\Phi}_{\tilde{r}_v+Nm,s}(t), \tag{8.11}$$

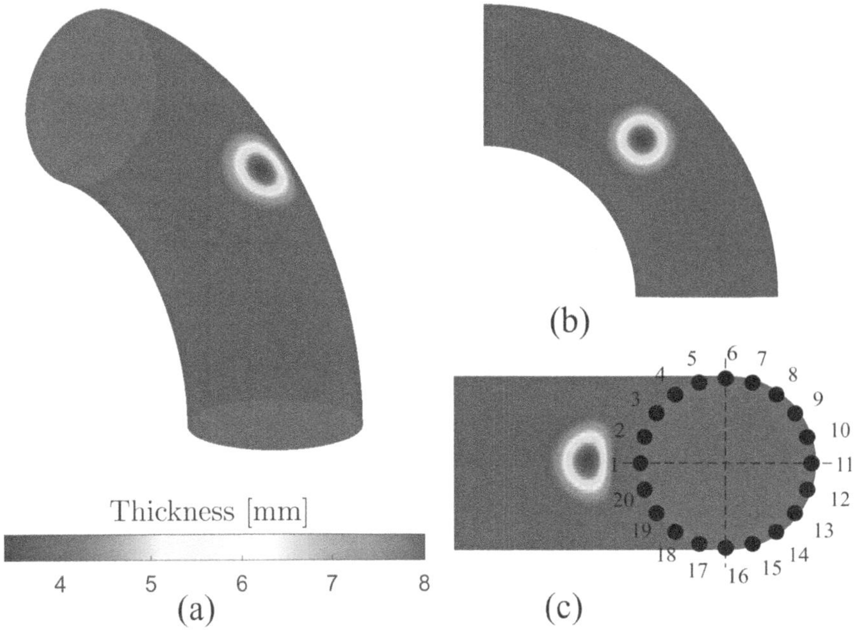

Figure 8.22 Hann-shaped defect with 50° thickness reduction located at 0° (a), 90° (b), and 180° (c) [184].

with $u^{hp}_{\tilde{r}_v,s}(t)$ being the helical-path-separated time traces. Note that the precision of Equations (8.9) and (8.10) depends on the distance traveled by the ray $x_{\tilde{r}_v,s}$. Hence, computing the shortest path $x_{\tilde{r}_v,s}$ is crucial for a successful implementation. Further details and this method's application to pipe bends can be found in [182].

8.3.4 RESULTS

8.3.4.1 Numerical study

Numerical investigation was conducted to evaluate the efficacy of FWI in reconstructing the thickness map of a pipe bend in the presence of Hann-shaped defects in different circumferential positions. The width of the defect was set at 120 mm for all cases, while the depth of the defect varied from 10% to 50%; it was placed in seven different circumferential positions α: 0°, 30°, 60°, 90°, 120°, 150°, and 180°, respectively. Figure 8.22 shows three main locations of the position of the defect: extrados (a), top (b), and intrados (c). The distribution of the transducers, for both numerical and experimental studies, is illustrated in Figure 8.22(c).

To quantitatively evaluate the reconstruction of the wall thickness loss map, we used the mean magnitude of relative error (MMRE) of the residuals between the true defect T and the reconstructed wall thickness maps Γ. The MMRE is the mean of the ratios of the absolute difference between two values to the magnitude of one of

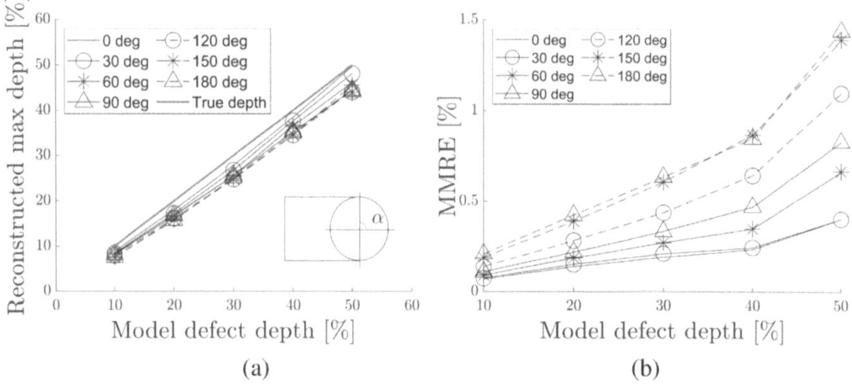

Figure 8.23 Maximum thickness maps reconstructed as a function of defect depth and position around the pipe circumference (a) and relative mean error for each case (b) [184].

the values. For a total of n ratios,

$$MMRE = \frac{1}{n} \sum_{i=1}^{n} \frac{|T_i - \Gamma_i|}{T_i}. \tag{8.12}$$

Figure 8.23(a) illustrates the relationship between the reconstructed maximum depth and the nominal depth of the defect, which varies with the position of the defect around the circumference of the pipe. The depth is expressed as a percentage of the nominal wall thickness. Similarly, Figure 8.23(b) shows how MMRE varies with the depth and position of the defect around the circumference.

Figure 8.23(a) makes it clear that for each circumferential position, the maximum reconstruction depth accuracy has a linear trend. In addition, the MMRE from Figure 8.23(b) shows a dependency on the circumferential position. The closer the defect is to the center of the 2-D domain, the more accurate the reconstruction.

Compared to GWT based on the TOF for a pipe bend [25], it is evident that FWI is capable of locating the defect regardless of its circumferential position. In TOF approaches, the ray paths in the bend avoid the extrados positions due to the focusing effect [25], limiting the coverage of the view angles. As a result, Brath et al. [24] proposed adding an additional array of transducers along the extrados position. In contrast, FWI takes advantage of the focusing effect, resulting in a more accurate reconstruction when the defect is located close to the extrados, as illustrated in Figure 8.23(b). In addition, since FWI does not rely on wave paths for the inversion scheme, a conventional pair of transducers is sufficient for the monitoring process.

8.3.4.2 Experimental validation

A smooth defect with a diameter of 100 mm and a thickness reduction of 47% was created in a bend with an angle grinder, as shown in Figure 8.24(a). The center of

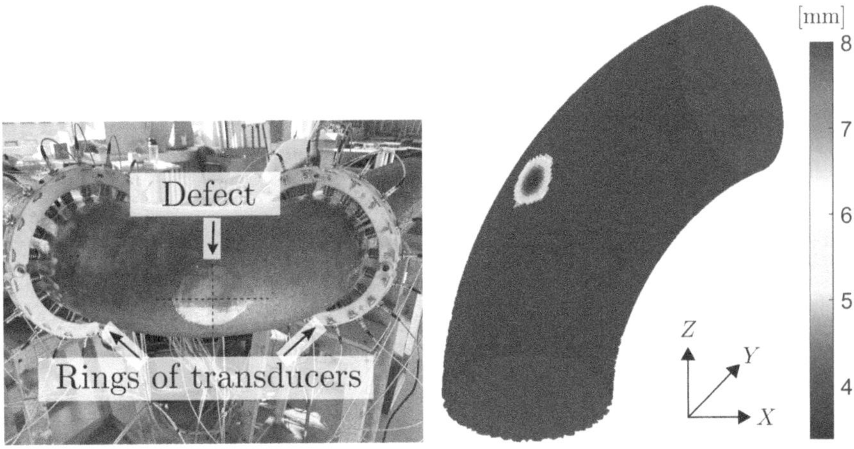

(a) (b)

Figure 8.24 Thickness reduction on a steel pipe bend (a) and its remaining thickness (b) [184].

the defect is at $[0,\pi/2]$, which are the circumferential and azimuth coordinates, respectively. Subsequently, Creaform's ACADEMIA 50 scanner was used to measure the remaining surface of the pipe bend. The scanner uses structured white-light technology, which involves projecting structured light onto an object and using cameras to capture the deformation of the light pattern. These data were then processed to generate a detailed 3-D model of the pipe bend, as shown in Figure 8.24(b).

The measurement setup is shown in Figure 8.24(a). It includes two rings, each containing 20 piezoelectric transducers with a resonant frequency of 43 kHz. These transducers, produced by Doppler, Ltd., in Guangzhou, China, are placed at both ends of the bend. The setup is complemented by a multiplexer for signal excitation and a data acquisition box. The transducers, which are uniformly distributed around the circumference, are spring-loaded against the pipe to facilitate radial excitation.

We measured the transmission of A_0 from each transducer to the opposite ring, resulting in a total of 400 recorded signals per transducer ring. The excitation source, the 2-D acoustic model, and the measurement response are shown in Figure 8.25. Figure 8.25(a) illustrates the excited five-cycle tone burst with a central frequency of 50 kHz, along with the corresponding response from one transducer ring when it is excited at the extrados position. The measured data are shown in Figure 8.25(b). Both wavefields have been replicated three times in accordance with the helical path separation algorithm. For reference, the wavefield without flaws, resulting from the 2-D acoustic forward model, is included in Figure 8.25(c). This model naturally does not include any helical paths because of the absorbing boundary conditions in the 2-D acoustic forward model. In contrast, Figure 8.25(d) displays the extracted wavefield after computing the helical path separation for the measured data.

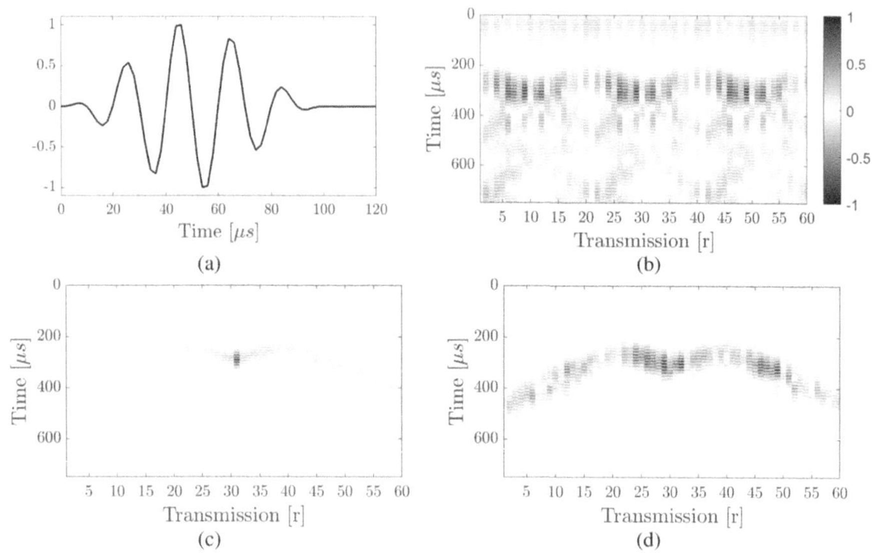

Figure 8.25 Excitation source (a), the recorded wavefields when excited at position 11 for the measurements (b), 2-D acoustic wavefield without a flaw (c), the helical-path-separated wavefields for the measurements (d) [184].

Figure 8.26(a) and (b) illustrate the 2-D representation of the real thickness reduction measured with the Creaform scanner and its reconstruction of the thickness at 50 kHz, obtained 40 iterations for the 3-D measured data, respectively. To further illustrate the construction of the thickness maps, Figure 8.26(c) and (d) show the axial profiles along the circumferential and azimuthal directions, respectively. For simplicity, the higher artifacts were removed. It can be seen that the length of the defect is significantly underestimated in the circumferential direction, as shown in Figure 8.26(c). However, Figure 8.26(d) shows good agreement in the azimuthal direction, resulting in an MMRE of 0.7.

The inaccuracy in the circumferential direction and the strong artifacts in the azimuthal direction can be attributed to the normalization of the scattered signals [185]. Moreover, the experimental results can be affected by the manufacturing tolerances of the bend, noise, the positioning of the transducers, errors in the helical path algorithm, coupling, and the performance of the transducer.

8.4 SUMMARY

This chapter presents an extension of UGWT using FWI to accurately determine Young's modulus, a crucial parameter for characterizing material stiffness in various fields such as medicine and structural mechanics. The method is based on the velocity reconstruction tomography which can be linked to Young's modulus through dispersion relations of Lamb modes. The simulation and experimental results demonstrate

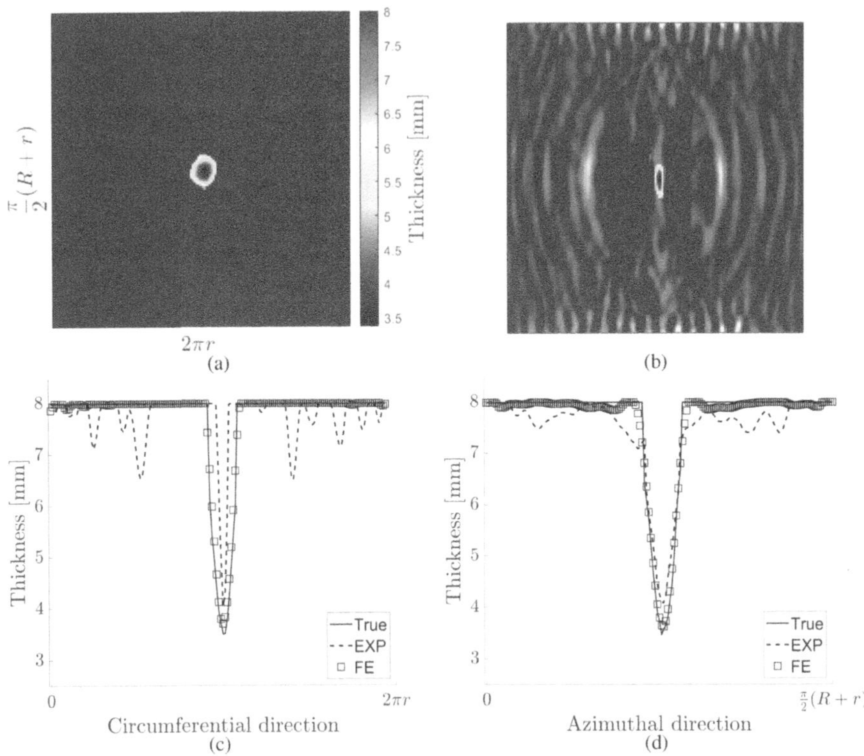

Figure 8.26 Real thickness map (a), the measured data (b), axial thickness profiles along the circumferential direction (c), and the azimuthal direction (d) [184].

the capability to reconstruct averaged through-thickness Young's modulus reductions, even with varying complexities, albeit with a slight reduction in resolution and accuracy compared to the theoretical limit.

To reconstruct stiffness defects in plates made of anisotropic materials, an approximate acoustic forward model was established based on Thompsen parameters that describe the angular variability of the velocity in the plate. The simulation results from FWI showed that the approach is very accurate in characterizing localized stiffness loss in weakly anisotropic plates. However, the method exhibits reduced accuracy when applied to highly anisotropic plates and becomes more challenging in characterizing defects with varying stiffness. In general, this method has significant potential to evaluate structures with varying stiffness, providing valuable information for a range of industries.

In addition to advances in stiffness defect characterization, this study also explores the application of UGWT for a pipe bend. The primary contribution is the demonstration of a 2-D acoustic model, which accounts for anisotropy through Thomsen parameters, enabling effective detection of wall thickness losses in pipe bends.

Using FWI and a smooth inversion process, the method successfully mapped defects in a carbon steel pipe elbow. Despite some limitations, such as slight underestimation of the depth of the defect, the method demonstrated its ability to locate defects and accurately detect reductions in wall thickness, even for defects located at various points along the bend. Compared to conventional travel-time-based tomography, this approach offers the advantage of mapping defects without being constrained by circumferential location, making it a promising tool for assessing the integrity of pipe bends in industrial applications.

8.5 EXERCISES

1. What is Young's modulus, and why is its accurate determination important?
2. What are the limitations of the pulse-echo method for evaluating Young's modulus?
3. What is the major challenge in estimating elastic properties using Lamb modes?
4. How are the dispersion curves of Lamb modes used to determine Young's modulus?
5. What are the limitations of conventional techniques assuming non-constant material properties along the wave propagation path?
6. What is the effect of Poisson's ratio on the phase velocity of guided waves at low frequencies?
7. What challenges arise when using guided wave testing for anisotropic waveguides?
8. How does the introduction of additional parameters into the model impact the inversion problem in UGWT?
9. How can a guided wave in a composite plate be related to an acoustic wave in a 2D medium?
10. What are the approximated Thomsen parameters used in the guided wave model?
11. Why does the solution of the acoustic equation become unstable when $\varepsilon < \delta$?
12. How can velocity and anisotropic parameters be derived in acoustic media?
13. What factors contribute to a decrease in the reconstruction accuracy in anisotropic materials?
14. Why does the accuracy of velocity reconstruction decrease when Thomsen parameters of defects diverge from background parameters?
15. What limitations are mentioned in the study regarding the assumptions made in the model and the impact of anisotropic parameters in different composite lay-up configurations?
16. How does FWI improve the accuracy of corrosion mapping in pipe bends compared to traditional travel-time tomography methods?
17. What is the significance of mapping bend's 3-D space coordinates directly into the 2-D domain?
18. How does the anisotropy of the wavefield change across the replicas of the pipe bend in the acoustic model?

19. What is the purpose of the helical path separation algorithm for a pipe bend?
20. What limitation of TOF approaches is mentioned in relation to the extrados position in the pipe bend, and how does FWI address this limitation?
21. What potential sources of error are mentioned that could affect the experimental results in pipe bend thickness reconstruction?

9 Tomography Based on Improved FWI

9.1 INTRODUCTION

Additive manufacturing is a rapidly growing manufacturing process, which is available to fabricate critical metal parts for aerospace, nuclear, and automotive engineering applications [2, 71, 223, 242, 263]. Compared to traditional manufacturing methods, additive manufacturing provides numerous benefits, including the ability to create complex geometries, enhance mechanical performance, and produce multi-material (MM) metallic components with tailored properties at specific locations [232, 243]. MM additively manufactured (AM) processes could make a superior bonding between multiple metals, as there are no weld seams. This enables the production of components that can withstand severe operating environments [153]. Consequently, ensuring the quality of the finished product is of utmost importance. Corrosion and wear are the main reasons for preferentially varying material properties within components. Additionally, the selective addition of desirable material properties during the fabrication to areas prone to damage can reduce the cost of creating components [264]. MM printed components are shown in areas where metal combinations provide benefits in properties that are superior to those obtained from the use of monolithic metal.

MM metallic components can be produced using powder bed fusion (PBF) and directed energy deposition (DED). The PBF method uses a high-energy power source to selectively melt or sinter a metallic powder bed [77,269,273]. Although MM parts fabricated by PBF have high geometrical accuracy [43, 54, 268], the process has a relatively low deposition rate and high cost. On the other hand, the DED method operates by feeding metal powders or wires into a melt pool generated by a heat source [82, 202]. The shapes are then created by depositing material layer-by-layer. DED allows MM manufactured components by changing the powders which are deposited at a specific time during the build. With multiple powder hoppers, using multiple materials can be as easy as turning off one hopper and another on. The deposited powders can be changed at any time, and different powders can be deposited in a single stream. Due to the relative ease of mixing and switching among different materials in real-time processing, MM printed structures in user-definable positions in which site-specific functionality is desirable can be created. Therefore, DED has been frequently used to produce MM printed parts [14, 32, 104].

Laser metal deposition (LMD) is a specific form of directed energy deposition (DED) that utilizes a laser as the energy source for metal deposition [118]. As LMD is a non-equilibrium processing technique [50], control of major parameters in LMD, including laser power, laser travel speed, laser spot size, layer thickness, step size of the gradient, and powder feed rate, is crucial in the control of MM printed

DOI: 10.1201/9781003667926-9

component's properties and quality. Shah et al. [209] found that the tensile strength of the MM component of Stainless Steel 316L and Inconel 718 is inversely proportional to the laser power and increases with the powder mass flow, rate. Mahamood et al. [128] investigated the optimized process parameters such as laser power, powder flow, and gas flow rate, for the Ti6Al4V/TiC composite, which was characterized through metallurgical, mechanical, and tribological studies. The results showed that the LMD built part has improved properties compared to the part produced at the constant process parameters. Amine et al. [4] related different cooling rates to the varied grain size, hence resulting in varied hardness and tensile strength of the multi-layered LMD parts. Reichardt et al. [190] found that the increased laser power and decreased beam travel speed can reduce the prevalence of powder inclusions and improve the mechanical properties of the gradient metal components.

Detrimental defects in MM LMD parts, such as porosities and inclusions, have been proven to be important contributors to a structure's propensity to rupture by serving as the crack growth location and the stress concentrator. For example, porosities have an influence on the overall LMD build quality because the conjoining of adjacent discontinuations severely affects mechanical properties (e.g., strengths and fatigue performance) [76, 110, 233]. NDE of defects plays an important role in the quality control of MM LMD parts. X-ray computed tomography (XCT) is one of the most commonly used NDE methods that use 2D radiographs to achieve a 3D outcome. It has shown good detection of the size, shape, and distribution of defects in AM components. However, the size of the test object that can be detected with XCT is small due to the limitations of XCT source power required for X-rays to penetrate the test object and the need for the XCT chamber to be large enough to contain the test object [228]. Also, defect detection based on XCT images is subject to high cost and the selected size threshold; for example, a larger threshold prevents inspection of smaller defects [226]. In addition, a non-contact NDE technique—thermographic technique—relies on temperature variations which indicate the existence of defects leading to the obstruction of heat conductivity in the metallic AM parts [15]. This technique is sensitive to surface features and useful for detecting larger areas. However, detecting deep defects is a challenging task for thermography.

In this chapter, we focus on ultrasonic phased array imaging methods in NDE, which can offer solutions to the limitations mentioned above. The common methods based on ultrasonic time-of-flight or ultrasonic attenuation have been used to detect defects in AM samples [85, 140, 224]. For example, the standard total focusing method which only considers the direct ray path of the ultrasound has been used to detect side-drilled holes contained in DED manufactured components [38]. However, most of the work has focused on the detection and characterization of single-layered AM components, with little attention being paid to MM printed parts. At the same time, not all useful information contained in ultrasonic signals is taken into account in the imaging methods for defect detection. As a way around such difficulties, advanced ultrasonic phased array imaging methods involve full acoustic or elastic wave equation modeling. Two popular methods are the reverse time migration (RTM) [176] and FWI [81]. The RTM based on cross-correlation of forward wavefields and time-reversed backward wavefields has shown its effectiveness in detecting and characterizing defects in AM printed components [180]. The advantages of the RTM are that no event picking is required and the imaging condition is easy to carry out. However, a key aspect of this method is a good starting model, especially when applied to real data [192].

The FWI that reconstructs model parameters (e.g., velocity) by minimizing the difference between observed data and modeled data has also demonstrated promise for ultrasonic NDE of defects [81, 148, 170, 208]. This method allows the full information contents of the ultrasonic data to be considered, and thus it is a powerful tool to obtain high-resolution reconstructions. However, the successful application of the conventional FWI method to MM or complex structures is dependent on a good starting model and the low-frequency information in the ultrasonic signals. When the modeled data are more than half a cycle away from the observed data, the cycle-skipping problem occurs, leading the inversion to converge to local minima. This results in the uninformative estimations of the model parameters [249]. The gradient sampling algorithm (GSA) is a potential alternative to tackle this challenge [30, 31], which was introduced to the conventional FWI in geophysics and verified by numerical simulations [75, 138, 267]. The advantage of GSA for the conventional FWI is to work with a set of perturbed velocity models near the current velocity model estimation (i.e., local neighborhoods) instead of a single velocity model. Therefore, the global information about the objective can be obtained by local gradient information, making it less sensitive to cycle skipping.

The objective of this chapter is to develop a two-stage ultrasonic phased array imaging method that combines the gradient sampling FWI (GSFWI) and the RTM for detecting and characterizing defects induced by lowering the laser power in the MM LMD parts of Inconel 625 (IN625) and 15-5PH Stainless Steel (PH15-5 SS). The low-frequency GSFWI is used to reconstruct the large-scale features (e.g., interface) of the unknown background and then this result is used as the starting model for the high-frequency RTM algorithm to obtain the reflectivity image of small-scale defects hidden in the MM LMD parts. This chapter is organized as follows. In Section 9.2, the principles of the GSFWI and the RTM are introduced, and how they can be combined to reconstruct MM LMD parts is described. The experimental procedure is presented in Section 9.3. In Section 9.4, the performance of the two-stage workflow for imaging interface and defects in the LMD-created MM specimens, especially in the low laser power zone, is analyzed, and then the results based on the two-stage method and XCT are cross checked. The discussion of this two-stage method is followed in Section 9.5. Finally, conclusions are summarized in Section 9.6.

9.2 METHODOLOGY

9.2.1 GRADIENT SAMPLING FWI

FWI is a data-fitting procedure in which the misfit between the modeled data d_{cal} and the observed data d_{obs} is minimized based on the time-domain or frequency-domain [234, 249]:

$$\min_{\mathbf{m}} f(\mathbf{m}) = \frac{1}{2}\|d_{cal}(\mathbf{m}) - d_{obs}\|_2^2, \tag{9.1}$$

where f is the nonlinear operator and $\|\cdot\|_2$ represents the L_2 norm. d_{cal} is computed in the current velocity model $\mathbf{m}$ based on 2D acoustic wave equation in heterogeneous media. Note that without loss of generality, the descriptions of the FWI and the RTM in this chapter are restricted to 2D problems. Most of the FWI methods are based on the local gradient-based optimization method to solve Equation (9.1) by iteration according to

$$\mathbf{m}^{(k+1)} = \mathbf{m}^{(k)} - \alpha^{(k)}\mathbf{G}^{(k)}, \tag{9.2}$$

where $\mathbf{m}^{(k)}$ is the updated velocity model at the kth iteration, α denotes the step length, and $\mathbf{G}$ is the search direction. If the steepest descent algorithm is used, the search direction is the gradient of the misfit function (Equation (9.1)). The gradient $\mathbf{G}(\mathbf{x})$ can be obtained by taking the partial derivative of the inverse wave equation with respect to $\mathbf{m}$ and applying the chain rule [234]

$$\mathbf{G}(\mathbf{x}) = 2\mathbf{m}(\mathbf{x}) \int U(\mathbf{x}_s, \mathbf{x}, t; \mathbf{m}) V(\mathbf{x}_r, \mathbf{x}, t; \mathbf{m}) d\mathbf{x}_s d\mathbf{x}_r dt, \tag{9.3}$$

where $\mathbf{x}_s$ and $\mathbf{x}_r$ are source and receiver positions, respectively. t is the time, and $\mathbf{x}$ is the grid point position. $U(\mathbf{x}_s, \mathbf{x}, t; \mathbf{m})$ is the source wavefields, and $V(\mathbf{x}_r, \mathbf{x}, t; \mathbf{m})$ is the receiver wavefields by back-propagating the data residual $\Delta d = d_{cal}(\mathbf{m}) - d_{obs}$.

Since the optimization problem as shown in Equation (9.1) is non-convex, using the vicinity of the global minimizer of the misfit function can help the gradient-based method to converge [124]. Therefore, the conventional FWI is dependent on the accuracy of the starting model and low-frequency information in observed data. To alleviate the dependence, we introduce the gradient sampling algorithm (GSA) into the FWI, named GSFWI, for the detection of MM LMD parts in this work. In each iteration of the GSFWI, the local gradient is calculated by sampling $N+1$ vectors $\mathbf{m}_i^{(k)}$ in a region that satisfies the condition $\left\|\mathbf{m}^{(k)} - \mathbf{m}_i^{(k)}\right\|_2^2 < \delta^{(k)}$, where $\delta^{(k)}$ is the maximum distance between the current velocity model $\mathbf{m}^{(k)}$ and a sampled one $\mathbf{m}_i^{(k)}$, i.e., the radius of the sampling region, as shown in Figure 9.1. The search direction is calculated as a weighted sum of all sampled gradients. Compared with the search direction of the conventional FWI based on the current velocity model estimation, the GSA has the ability to search the local neighborhoods of the current velocity model estimation, and thus it has a higher chance of finding a better search direction. However, this procedure requires the calculation of gradients at many sampling points within the sampling region (neighborhoods of the current velocity model). To reduce the computational cost, we assume that the influence of the small perturbation velocity model near the current velocity model can be approximated by time shift or space shift [124, 267]. In this chapter, we can approximate source and receiver wavefields corresponding to the slightly perturbed velocity models near the current model

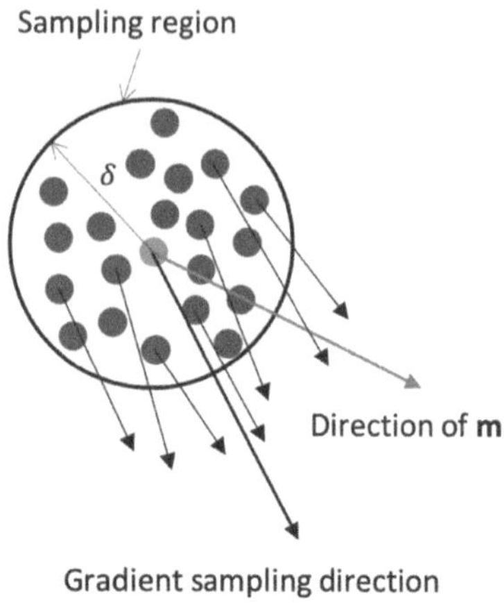

Figure 9.1 Gradient sampling. The green dot represents the current velocity model estimation and the purple dots denote perturbed velocity models around the current velocity model estimation [177].

estimation by the space shifting wavefields. Therefore, the gradient of the GSFWI can be expressed as [267]:

$$\mathbf{G}_{GSA}(\mathbf{x}) = 2\mathbf{m}(\mathbf{x}) \int U(\mathbf{x}_s, \mathbf{x} - \mathbf{h}_{t\mathbf{x}}, t; \mathbf{m}) V(\mathbf{x}_r, \mathbf{x} - \mathbf{h}_{t\mathbf{x}}, t; \mathbf{m}) d\mathbf{x}_s d\mathbf{x}_r dt d\mathbf{h}_{t\mathbf{x}}, \quad (9.4)$$

where $\mathbf{h}_{t\mathbf{x}}$ is the random space shift at each time step per model grid. The shift range should be proportional to the propagation distance and the velocity error, and the direction of the space shift is consistent with the direction of the wave propagation. From early iterations to later iterations of the inversion, as the velocity model is more accurate, the space shift is smaller. Note that the GSFWI (see Equation (9.4)) has the same computational cost as the conventional FWI (see Equation (9.3)). Also, the GSFWI can provide a good starting model for the conventional FWI in which low-frequency information is required to be implemented.

9.2.2 REVERSE TIME MIGRATION

RTM reconstructs defects in a structure by cross-correlating the forward and backward wavefields [176, 178, 277]. The imaging condition between both wavefields is given by

$$I(\mathbf{x}) = \sum_{sources} \int_0^T W_s(\mathbf{x}, t) W_r(\mathbf{x}, T - t) dt, \quad (9.5)$$

where $I(\mathbf{x})$ is the reflectivity image and T is the time duration. $W_s(\mathbf{x}, t)$ is the forward wavefields which are obtained by applying the excited signals at source positions. $W_r(\mathbf{x}, T - t)$ is the backward wavefields, which means that the observed data (measured data) at receivers are time reversed. For the RTM method to perform successfully, the background velocity model is assumed to be accurate enough. An incorrect

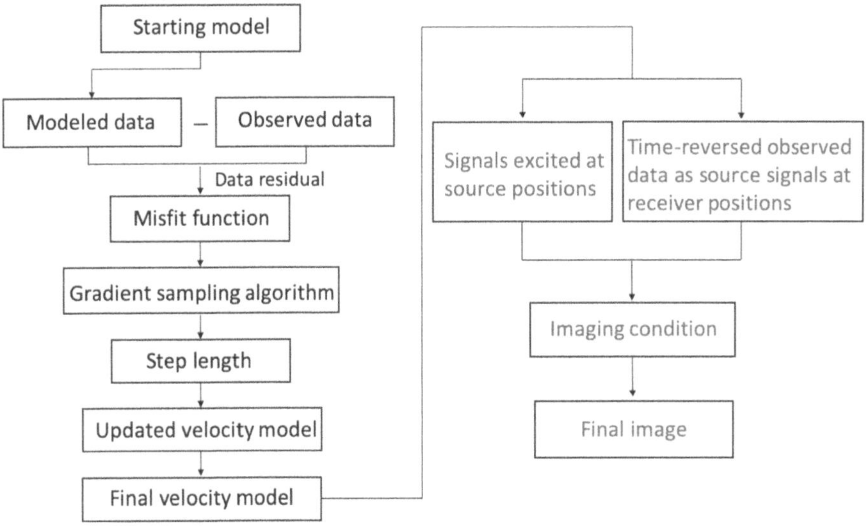

Figure 9.2 Workflow of the two-stage method. The GSFWI provides an accurate background velocity model which is then used as the starting model for the RTM targeting small-scale defects [177].

velocity model can result in an unfocused reflectivity image or even spurious reflectors [148, 258]. Note that the receiver wavefields mentioned in FWI/GSFWI refer to the time reversal of the data residual.

9.2.3 COMBINED GSFWI AND RTM WORKFLOW

Detection and characterization of interfaces and defects in MM LMD parts is challenging in ultrasonic phased array imaging. In this work, a new two-stage method combining GSFWI and RTM is proposed. Because inaccurate velocity model reconstructions can cause problems in imaging applications. Therefore, the GSFWI is used to correct anomalies in the ultrasonic wave velocity followed by a high frequency RTM to reconstruct small-scale features. Figure 9.2 outlines the workflow of this two-stage method. Next, the performance of this two-stage method through two experiments of MM LMD parts is demonstrated.

9.3 EXPERIMENTAL PROCEDURE

9.3.1 LASER METAL DEPOSITION PROCESS

Inconel 625 (IN625) and 15-5PH Stainless Steel (PH15-5 SS) powder feed materials with a 45-105 μm particle size range were employed as feedstock materials. The material properties are shown in Table 9.1. LaserTec65 3D hybrid was applied to

Table 9.1
Material properties [177]

Material	Young's modulus (GPa)	Poisson's ratio	Density (kg/m^3)
IN625	200	0.294	8220
PH15-5 SS	196	0.272	7800

Table 9.2
LMD process parameters employed in Specimen 1 of IN625 and PH15-5 SS [177]

	IN625 zone	PH15-5 SS zone	Lower laser power zone
Laser power (W)	1700	1700	600
Powder flow rate (g/min)	15	15	15
Gas flow rate (L/min)	6	6	6
Traverse speed (m/min)	1	1	1
Layer thickness in the z direction (mm)	0.1	0.1	0.1

produce the MM specimens using the laser metal deposition (LMD) process.

A diode laser was used as a laser source and argon gas as the shielding gas. Dimensions of Specimen 1 are shown in Figures 9.3(a). The LMD process parameters employed in this build are shown in Table 9.2. To study the influence of process parameters on material porosity, the zone with the sizes of 15 mm×15 mm×6 mm of Specimen 1 was fabricated by lowering the laser power from normal 1700 W to 600 W. Specimen 2 of an IN625 block component (27 mm×20 mm×9 mm) and a PH15-5 SS block component (27 mm×20 mm×9 mm) was fabricated by varying the laser power, as shown in Figure 9.3(b). In Specimen 2, an artificial defect with the sizes of 2 mm×20 mm×1 mm was introduced by lowering the laser power to 0 W. The laser power on both sides of this defect was 800 W ($z \in [9$ mm, 10 mm$]$). Note that all the remaining zones of Specimen 2 were fabricated with the laser power of 1500 W. The powder flow rate was 15 g/min, the gas flow rate was 6 L/min, and the traverse speed was 1 m/min. Two specimens were sand blasted before the detection to obtain a uniform surface for better signal reception.

9.3.2 ULTRASONIC TESTING

The experimental setup for ultrasonic testing is shown in Figure 9.4. A 64-element ultrasonic linear array manufactured by Guangzhou Doppler Electronic Technologies Co., Ltd. was used to excite and receive signals. The linear array with a central frequency of 5 MHz and with an element pitch of 0.6 mm was placed on the top

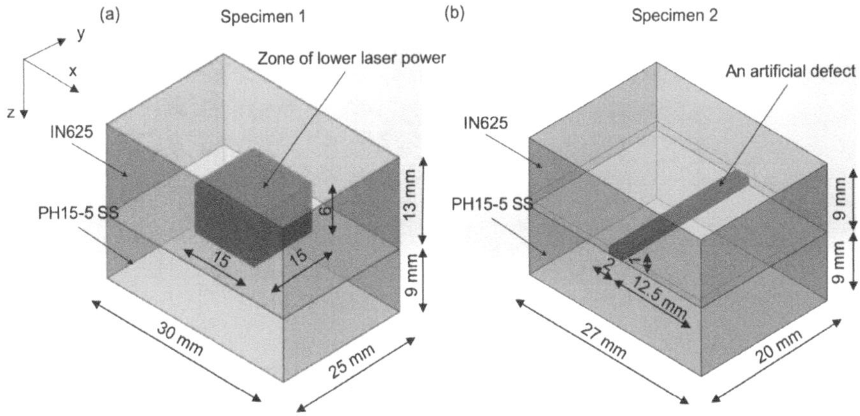

Figure 9.3 (a) Dimensions of Specimen 1 of IN625 and PH15-5 SS with high enough laser power at 1700 W and with the zone of reduction in laser power at 600 W. (b) Sizes of Specimen 2 with an artificial defect (laser power at 0 W) and with the reduction in laser power of 800 W on both sides of this defect (z∈[9mm, 10mm]). The remaining zones of Specimen 2 were fabricated by the laser power of 1500 W. Note that sizes are not to scale [177].

Figure 9.4 Experimental setup for ultrasonic testing [177].

surface of the MM LMD specimens. Each element transmits a signal in turn, which is then measured by all of the elements in the array. This produces a set of signals for all source-receiver combinations in the full matrix capture (FMC) data. Note that only 50×50 combinations in the FMC data are used for Specimen 1, and a 45×45 matrix is used for Specimen 2 due to the limitation of the specimen size in the x direction. Figure 9.5 shows the typical contour plot of the ultrasonic displacement signals measured over all receivers in the experiment.

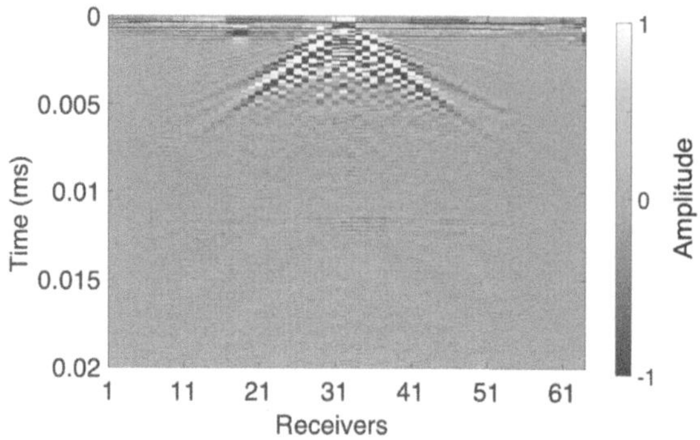

Figure 9.5 Typical contour plot of experimentally observed data in one of the array elements [177].

Table 9.3
XCT scan parameters [177]

Source voltage (kV)	428
Source current (μA)	200
Angle increment (°)	0.25
No. of projections	1440

9.3.3 XCT IMAGE SLICE ACQUISITION

The MM LMD specimens were scanned on a Nikon Metrology XTH 450 X-ray CT system. This system is equipped with 450 kV open tube micro-focus X-ray source and a flat panel detector with 2000×2000 pixels, where the pixel size is 200 μm. For each scan, the MM LMD specimen was illuminated by cone beam X-rays which were transmitted through the specimen and detected by the detector for the measurement of X-ray beam intensities. The specimen was rotated over 360° and a series of 2D images, i.e., projections, were obtained at the angular increments for the reconstruction of images. XCT scan parameters are shown in Table 9.3. The source voltage and current values were selected according to the thickness and materials of the specimens. A resolution with a voxel size of about 38 μm for CT scanning was used.

9.4 EXPERIMENTAL RESULTS

9.4.1 IMAGING OF SPECIMEN 1

In this work, the XCT image is presented first, which is used for cross-checking with results based on the proposed two-stage method. The reconstruction of XCT image

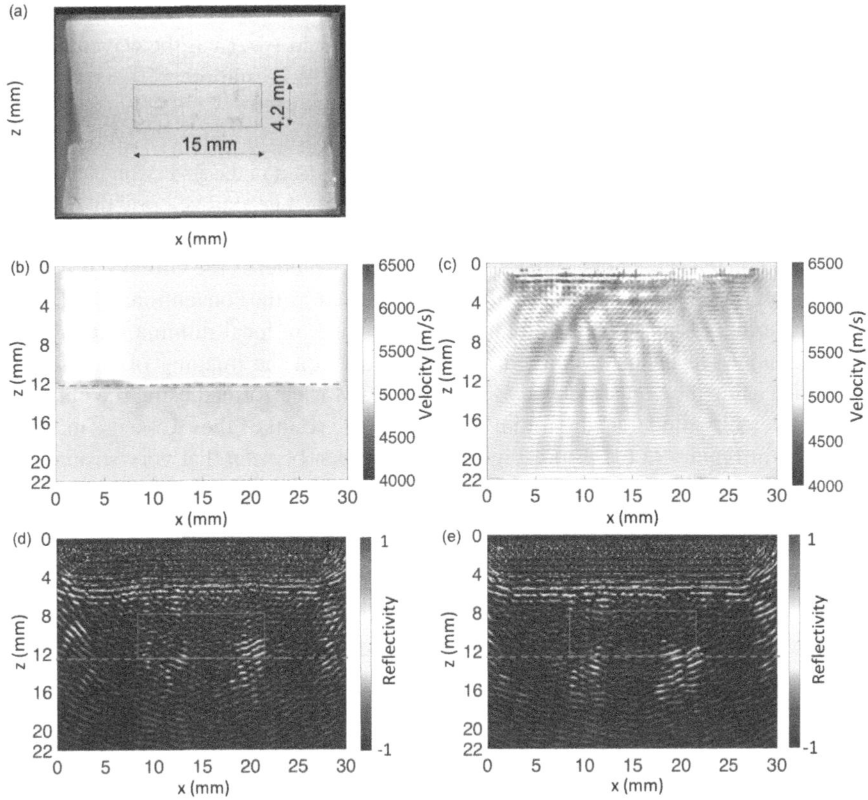

Figure 9.6 Reconstructions of Specimen 1 with the reduced laser power at 600 W in a specified zone (red solid region). (a) XCT image at y = 13.4 mm, (b) GSFWI image, (c) conventional FWI image, (d) RTM image using obtained GSFWI model as the initial model (i.e., the two-stage method), and (e) RTM image based on the homogeneous initial model [177].

slices of Specimen 1 with the zone of the reduced laser power is carried out by the open-source ImageJ/Fiji, which is used to characterize and analyze the internal structure of MM LMD specimens, allowing the good understanding of the effect of process parameters on material porosity. The reconstructed XCT image in the x-z plane at y = 13.4 mm, as shown in 9.6(a), is 2D eight-bit grayscale image, without physically cutting them open. It can be seen from Figure 9.6(a) that the porosities show darker than the solid material because of the differences in X-ray attenuation. The size of the defect area in the x-z plane is 15 mm × 4.2 mm in the XCT image. Note that the size of the defect area in different x-z planes along the y-axis of the laser power reduction zone is basically the same.

In the proposed two-stage method, the GSFWI algorithm is first applied to the experimental FMC data from Specimen 1. The homogeneous starting model with a longitudinal wave (L-wave) velocity of 5770 m/s is used in the inversion. In the GS-FWI, the range of the random space shift is from -18λ to 18λ (λ is the wavelength at 6 MHz, i.e., 0.962 mm). Figure 9.6(b) shows the GSFWI result after 20 iterations. It can be seen that the interface between the IN625 layer and PH15-5 SS layer is clearly reconstructed, which shows good consistency with the true model (see Figure 9.3(a)). Because the longer wavelength (the lower frequency) used in the inversion is targeted at the reconstruction of the large-scale feature of the background velocity model. For comparison, Figure 9.6(c) shows the reconstruction obtained from the conventional FWI based on the same homogeneous initial velocity model as the GSFWI. It is clear that the interface cannot be successfully reconstructed by the conventional FWI. This is because the conventional FWI can easily get stuck in local minima in this case. The second stage in the combined method focuses on the imaging of small-scale porosities using the high-frequency RTM. The RTM is performed using the obtained GSFWI velocity model as the initial model, and the result of the two-stage method, is shown in Figure 9.6(d). From Figure 9.6(d), it can be seen that very strong artifacts appear near the positions of the ultrasonic phased array, which are caused by the near-field clutter [20]. Most of the porosities are randomly distributed and irregularly shaped in the region of the laser power at 600 W. This is due to the lower laser power, which leads to insufficient melting of the metal powder particles, resulting in their entrapment within porosities during rapid solidification. The accuracy of the reconstruction result of the two-stage method is cross-checked by cross-comparison with the XCT image (see Figure 9.6(a)), showing a good agreement.

Figure 9.6(e) shows the RTM image based on the homogeneous initial model with the L-wave velocity of 5770 m/s. As expected, the RTM based on the obtained GSFWI model as the initial model (i.e., two-stage method) offers more accurate reconstructions of porosities than the RTM based on the homogeneous initial model. Figure 9.6(e) shows the significant downward shift in the locations of the porosities compared to Figure 9.6(d). Because variations in L-wave velocity in the two different layers and the laser energy reduction region are not considered in the RTM based on the homogeneous initial velocity model. This validates the importance of an accurate initial model for the RTM.

9.4.2 IMAGING OF SPECIMEN 2

Specimen 2 considered in this chapter is an MM LMD part with an artificial defect (laser power at 0 W) just above the interface between the IN625 layer and PH15-5 SS layer and with the zones of lower laser power at 800 W on both sides of this defect. Figure 9.7(a) shows the XCT image reconstruction in the x-z plane at y = 8 mm. It is clear that the XCT image provides good detection and characterization of the location and geometry of the artificial defect. By lowering the laser power on both sides of the artificial defect ($z \in [9\text{ mm}, 10\text{ mm}]$), some porosities located on either side of this artificial defect are also observed.

Figure 9.7(b) shows the GSFWI image based on a homogeneous initial model (L-wave velocity of 5770 m/s). It can be seen that a good reconstruction of the interface between the IN625 layer and the PH15-5 SS layer is achieved. As a comparison, the

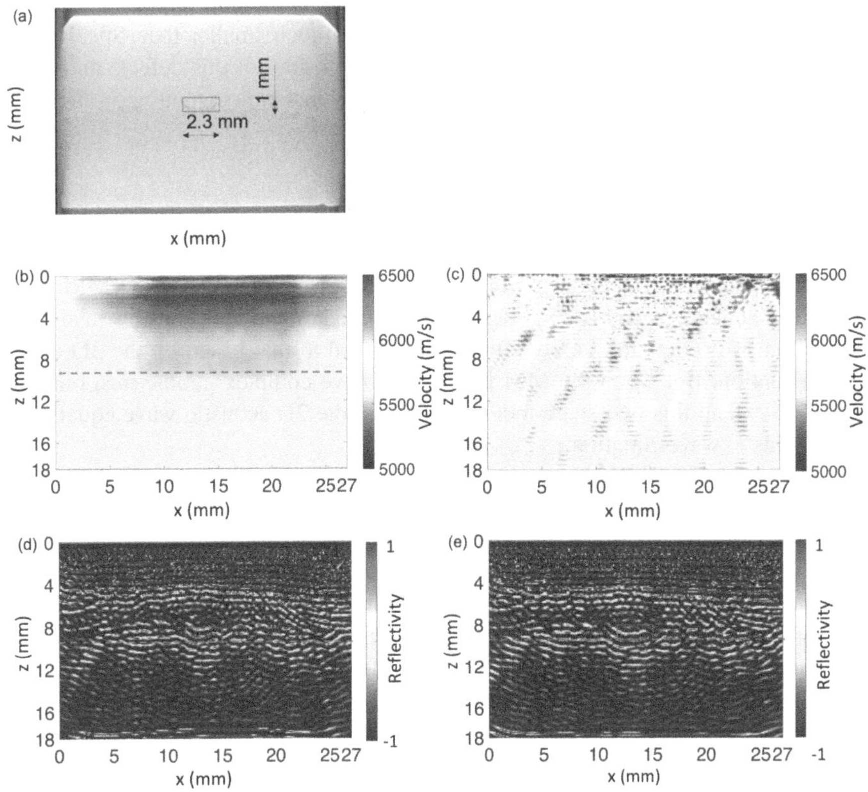

Figure 9.7 Reconstructions of Specimen 2 with an artificial defect (red solid region) just above the interface between IN625 and PH15-5 SS and with the reduced laser power zone (red dashed regions). (a) XCT image at y = 8 mm, (b) GSFWI image, (c) conventional FWI image, (d) RTM image using obtained GSFWI model as the initial model (i.e., the two-stage method), and (e) RTM image based on the homogeneous initial model [177].

result obtained from the conventional FWI based on the same homogeneous initial model is shown in Figure 9.7(c). It is obvious that the conventional FWI cannot accurately reconstruct the velocity model of Specimen 2. Figures 9.7(d) and (e) show the RTM reflective images based on the reconstructed background velocity model obtained from the GSFWI and the homogeneous initial velocity model, respectively. These two figures also show strong artifacts near the array position and the bottom of Specimen 2 due to the near-field clutter and reflections from the backwall, respectively. Compared to Figure 9.7(d) obtained from the two-stage method, Figure 9.7(e) shows that the location of the artificial defect is slightly offset from the true model (see Figure 9.3(b)) in the z-direction. It should be noted that the defect location deviation from the true model or the XCT image (see Figure 9.7(a)) is not as obvious as that in Specimen 1, because the range of the velocity variations in the z direction caused by laser power reduction in Specimen 2 is much smaller than Specimen 1. Besides the artificial defect, some porosities on both sides of this defect can also be observed in Figures 9.7(d) and (e). This is caused by incomplete melting of the metal powder at the low laser power of 800 W.

9.5 DISCUSSION

The results shown in this work have demonstrated that the two-stage method has the ability to achieve good reconstructions of interfaces and defects in the MM LMD parts. The advantage of this two-stage method based on the 2D acoustic wave equation is that it is computationally efficient compared to that based on the 3D elastic wave equation. But when the MM LMD parts have complex geometries, our next step is to extend this two-stage method based on the 2D acoustic wave equation to the 3D elastic wave equation.

The proposed GSFWI in the two-stage method can achieve accurate interface reconstruction based on a homogeneous initial model and without starting from low-frequency components in the experimental data. The GSFWI calculates sets of gradients at sampling points near the current velocity model (see Figure 9.1) rather than the single gradient computed in the conventional FWI. Therefore, it is capable of finding the minimum value of a non-convex minimization problem. However, its computational cost is high and it is not feasible in practice settings. To overcome this issue, we calculate the approximations of source and receiver wavefields corresponding to slight perturbation velocity models in the vicinity of the current velocity model estimation by space-shifting these wavefields. The range of the space-shift is obtained by trial and error in this work, and we will further develop adaptive space shifts to accelerate the convergence in future work.

9.6 SUMMARY

In this work, the two-stage ultrasonic waveform tomographic method is developed to detect and characterize defects in the MM LMD parts. In the first stage, the gradient sampling full waveform inversion (GSFWI) is proposed to reconstruct large-scale features of interest in MM LMD parts, which can mitigate the cycle-skipping issue by working with a set of velocity model estimations near the current model estimation rather than a single model estimation. Therefore, it has the ability to find the minimum value of the objective function and achieve accurate background velocity estimation. By taking the GSFWI reconstruction as a background model, the high-frequency reverse-time migration is then applied to image the small-scale defects. The experimental results show that the two-stage method has promise for reconstructing the interface and small porosities induced by lowering the laser power in MM LMD parts. In addition, the results show good agreement with XCT images.

9.7 EXERCISES

1. Compare the advantages and disadvantages of NDE methods such as XCT, thermographic technique, and ultrasonic method.
2. Compared with the detection of single-layered AM components, please analyze the difficulty of detecting MM printed parts.
3. Please describe the basic principles of reverse time migration (RTM) and full waveform inversion (FWI).
4. Why combine RTM and FWI? What is the purpose of RTM at low frequency and FWI at high frequency?
5. Compared with the conventional FWI using the steepest descent method, what are the characteristics and advantages of the gradient sampling algorithm (GSA) proposed in this chapter?
6. Please use a graphical description to draw the difference between the GSA and the steepest descent method.
7. Please draw a workflow for combining GSFWI and RTM.
8. Please describe the effect of different printing parameters, such as laser power and laser travel speed, on the quality of the printed products.
9. Please analyze the signals in Figure 9.5, which are longitudinal waves, which are transverse waves, and which are direct waves.
10. Please analyze why the conventional FWI cannot achieve interface reconstruction in Specimen 1.
11. In Figure 9.6, what is the difference between Figure 9.6(d) and Figure 9.6(e)? Please analyze the reason.
12. Please compare the results of Specimen 1 with those of Specimen 2, and analyze the reasons for their differences.
13. When the MM LMD parts have complex geometries, how to improve the algorithms to realize the detection of them?

10 Multi-parameter FWI

10.1 INTRODUCTION

Quality control plays a crucial role in the aviation and automotive industries, particularly for large forging components and powder metal compacts. Various parameters, including density, elastic moduli, and tensile strength, are important for assessing mechanical performance and ensuring quality assurance. Density is a key parameter that correlates with elastic moduli and strength, making it a critical factor in quality control. For example, increasing the density of powder metal components can improve their fatigue strength [274]. Fluctuations in density can significantly impact mechanical performance. In powder metallurgy, for instance, the powder flows into the die cavity at the early stage of the compaction. However, as the pressure increases and the movement of the powder becomes restricted, an uneven density distribution may occur. The existence of a low-density region in the internal corner of the part could prove to be a fatal flaw, as the corner is always the point of stress concentration when the part is in service [74]. Currently, standardized density tests involve gravimetric measurements on specimens cut from components [11], which are intrusive. Therefore, the development of NDE techniques that can determine the density distribution is necessary.

Ultrasonic testing is a widely used and cost-effective alternative for NDE of engineering components. When ultrasonic waves propagate through metal parts, valuable information about the material's physical properties, such as velocity, density, and attenuation, can be obtained by analyzing the characteristics of the waves [150]. Much of the research on characterizing density distribution relies on ultrasonic velocity measurements [86, 154]. The density of the sample is indirectly determined from the velocity of ultrasonic waves propagating through the sample, assuming Young's modulus and Poisson's ratio are known. However, given the close relationship between Young's modulus and density, achieving accurate density reconstruction can be challenging. Additionally, these methods are often time-consuming, particularly when applied to inaccessible regions.

Ultrasonic tomography offers significant potential for estimating the density distribution of metal parts without accessing all points on the surface. An ultrasonic tomography method has been utilized to determine the density variation of powder metal components, using a filtered back-projection algorithm based on the travel time of ultrasonic bulk L-waves [274]. The resulting velocity reconstruction lays the foundation for predicting the density distribution. However, this approach only reconstructs velocity, neglecting the coupled effects between different parameters. Furthermore, this method is only valid when inhomogeneities are much larger than the wavelength, as it overlooks diffraction effects [274].

In this chapter, a novel ultrasonic tomography method is developed to simultaneously quantify the density variation and velocity distribution of metal parts. The

DOI: 10.1201/9781003667926-10

reconstruction of velocity and density is achieved through multi-parameter FWI, which allows for the consideration of higher-order diffraction and scattering. Compared to the mono-parameter FWI, the multi-parameter FWI considers the density term in the 2D acoustic model in this work. However, transitioning from mono-parameter inversion to multi-parameter inversion introduces increased indeterminacy. Different parameters with distinct characteristics can exhibit coupled effects in the propagation region, ranging from transmission to reflection, leading to cross-talk between the parameters. Additionally, the disparities in amplitude among the various parameters in the wave field pose a challenge that needs to be properly scaled to prevent ill-conditioning in the inversion process [150].

To mitigate the effects of cross-talk and rescale the amplitudes of different parameters, a robust estimation of the inverse Hessian operator plays a very important role in the multi-parameter inversion [137,150]. In the multi-parameter inversion, the off-diagonal blocks of the Hessian operator provide the coupling information between different parameters. Therefore, an accurate estimation of the inverse Hessian in the inversion can mitigate the trade-off effect in the reconstruction. In practical applications, calculating or storing the inverse Hessian can be computationally challenging due to its large size. The approximation of inverse Hessian can be obtained using the limited memory Broyden-Fletcher-Goldfarb-Shanno method [149], but this method relies heavily on the initial model and iterative scheme [27]. Sheen [210] used low-rank approximation to decompose the Hessian in the time domain FWI. However, it requires a large memory capacity to store this Hessian. In this chapter, the truncated Gauss-Newton method is introduced to better estimate the inverse Hessian in multi-parameter FWI for simultaneous reconstruction of velocity and density. The approximate Hessian does not need to be stored in memory in advance [123].

In this chapter, the theory of multi-parameter FWI based on the truncated Gauss-Newton method is described in Section 10.2. In Section 10.3, simultaneous reconstruction of velocity and density from the simulated data is presented. Section 10.4 then applies the multi-parameter FWI to the real experimental data from a diffusion-bonded cylinder of brass/steel. The conclusions are summarized in Section 10.5.

10.2 THEORY

The frequency-domain 2D acoustic wave equation in the isotropic heterogeneous media can be expressed as

$$\frac{\omega^2}{\rho(x,y)v^2(x,y)}p(x,y,\omega)+\nabla\cdot\left(\frac{1}{\rho(x,y)}\nabla p(x,y,\omega)\right)=-s(x,y,\omega), \tag{10.1}$$

where $p(x,y,\omega)$ is the frequency-domain pressure (or displacement) wave-field and ω is the angular frequency. $\rho(x,y)$, $v(x,y)$, and $s(x,y,\omega)$ are the density, the velocity and the source, respectively.

The aim of multi-parameter FWI is to find the model m by minimizing the data residual Δd between the data $d_{calc}^{(k)}$ computed in the current model $\mathrm{m}^{(k)}$ and the observed data d_{obs}. A least-squares norm of the objective function is given by

$$C^{(k)}(\mathrm{m})=\frac{1}{2}\Delta\mathrm{d}^{\dagger}\Delta\mathrm{d}, \tag{10.2}$$

where $\Delta d = d_{calc}^{(k)} - d_{obs}$ is the data residual and k is the iteration number. The superscript † represents the transposed conjugate. In this chapter, the model m is defined as

$$\mathrm{m(x,y)} = [v(x,y) \quad \rho(x,y)]^T. \tag{10.3}$$

Local minimization of the misfit function is given by

$$G^{(k)}(\mathrm{m}) = \nabla C^{(k)}(\mathrm{m}) = -H^{(k)}(\mathrm{m})\Delta \mathrm{m}^{(k)}, \tag{10.4}$$

$$i.e., \quad \Delta \mathrm{m}^{(k)} = -(H^{(k)}(\mathrm{m}))^{-1} G^{(k)}(\mathrm{m}), \tag{10.5}$$

where Δm represents the descent direction. $G(\mathrm{m})$ and $H(\mathrm{m})$ are the first and second derivatives of the misfit function with respect to the model m, i.e., the gradient and the Hessian

$$G(\mathrm{m}) = \frac{\partial C}{\partial \mathrm{m}} = (\frac{\partial d_{calc}}{\partial \mathrm{m}})^{\dagger} \Delta \mathrm{d} = J^{\dagger} \Delta \mathrm{d}, \tag{10.6}$$

$$H(\mathrm{m}) = \frac{\partial^2 C}{\partial \mathrm{m}^2} = J^{\dagger} J + (\frac{\partial J}{\partial \mathrm{m}})^{\dagger} \Delta \mathrm{d}, \tag{10.7}$$

where J is the Fréchet derivative matrix (i.e., the sensitivity kernel). It needs to be mentioned that the second term in Equation (10.7) is usually difficult to calculate, and it has relatively small values [236]. Thus, this is commonly dropped off, and then the approximate Hessian H_a can be written as

$$H_a(\mathrm{m}) = J^{\dagger} J. \tag{10.8}$$

The model $\mathrm{m}^{(k+1)}$ is updated iteratively along the descent direction Δm according to

$$\mathrm{m}^{(k+1)} = \mathrm{m}^{(k)} + \alpha^{(k)} \Delta \mathrm{m}_k, \tag{10.9}$$

where α is a step length using the linear-search method and $\alpha \Delta$m is the perturbation model.

In this chapter, the descent direction Δm and kernel matrix J are given by

$$\Delta \mathrm{m} = \begin{pmatrix} \Delta v \\ \Delta \rho \end{pmatrix}, \tag{10.10}$$

$$J = (J_v \quad J_\rho), \tag{10.11}$$

where J_v and J_ρ are the sensitivity kernels of velocity and density. Therefore, the Gauss-Newton normal equation can be written as follows [266]

$$\begin{pmatrix} J_v^{\dagger} \\ J_\rho^{\dagger} \end{pmatrix} (J_v \quad J_\rho) \begin{pmatrix} \Delta v \\ \Delta \rho \end{pmatrix} = -\begin{pmatrix} J_v^{\dagger} \\ J_\rho^{\dagger} \end{pmatrix} \Delta \mathrm{d}, \tag{10.12}$$

where

$$\begin{pmatrix} J_v^\dagger \\ J_\rho^\dagger \end{pmatrix} \begin{pmatrix} J_v & J_\rho \end{pmatrix} = \begin{pmatrix} J_v^\dagger J_v & J_v^\dagger J_\rho \\ J_\rho^\dagger J_v & J_\rho^\dagger J_\rho \end{pmatrix}. \tag{10.13}$$

This Gauss-Newton approximate Hessian includes a 2×2 block matrix. The diagonal blocks represent the correlations of the sensitivity kernel J with parameters of the same nature. In each sub-matrix, the diagonal elements are calculated by the zero-lag auto-correlation of J at the same spatial position, which provide information on coincident-in-space model parameters of different natures. The off-diagonal blocks are the correlations of J with the parameters of different natures and thus reflect the effects of cross-talk between velocity and density. These off-diagonal elements in each sub-matrix are obtained from the zero-lag cross-correlation of J at different spatial positions, which contain the information of the interaction between parameters of different natures [266]. The inverse of Gauss-Newton approximate Hessian can be used to mitigate the trade-off effects and rescale the magnitudes of velocity and density.

The calculation of the misfit function gradient in Equation (10.6) is a key step in FWI, which can be carried out by the adjoint-state (AS) method [160, 234] or the scattering-internal (SI) method [40, 123, 204]. The AS approach used in Chapter 3 is based on cross-correlating the forward wave-field with the back propagated residual wave-field to obtain the gradient. In this chapter, the SI method proposed by Liu et al. [123] evaluates kernel-vector Products in Equation (10.6) by accumulating decomposed vector-scalar product without explicitly calculating the matrix (see Equation (10.14)). It needs to be noted that the SI method is a useful alternative, which is not a replacement for the AS method [266].

Assuming that the data residual $\Delta \mathrm{d}$ has a size of $l \times 1$ and the model parameters of velocity and density have the dimension of $n \times 1$, then kernel matrix and its conjugate transpose have the sizes of $l \times n \times 2$ and $2 \times n \times l$, respectively. The gradient of the misfit function in Equation (10.6) can be given by:

$$G = J_m^\dagger \Delta \mathrm{d} = \tag{10.14}$$

$$\begin{pmatrix} \overline{J_{v11}} \cdots \overline{J_{vi1}} \cdots \overline{J_{vl1}} \\ \vdots \quad \ddots \quad \vdots \quad \ddots \quad \vdots \\ \overline{J_{v1j}} \cdots \overline{J_{vij}} \cdots \overline{J_{vlj}} \\ \vdots \quad \ddots \quad \vdots \quad \ddots \quad \vdots \\ \overline{J_{v1n}} \cdots \overline{J_{vin}} \cdots \overline{J_{vln}} \\ \overline{J_{\rho 11}} \cdots \overline{J_{\rho i1}} \cdots \overline{J_{\rho l1}} \\ \vdots \quad \ddots \quad \vdots \quad \ddots \quad \vdots \\ \overline{J_{\rho 1j}} \cdots \overline{J_{\rho ij}} \cdots \overline{J_{\rho lj}} \\ \vdots \quad \ddots \quad \vdots \quad \ddots \quad \vdots \\ \overline{J_{\rho 1n}} \cdots \overline{J_{\rho in}} \cdots \overline{J_{\rho ln}} \end{pmatrix} \begin{pmatrix} \Delta \mathrm{d}_1 \\ \vdots \\ \Delta \mathrm{d}_\mathrm{i} \\ \vdots \\ \Delta \mathrm{d}_\mathrm{l} \end{pmatrix} = \sum_{i=1}^{l} \begin{pmatrix} \overline{J_{vi1}} \\ \vdots \\ \overline{J_{vij}} \\ \vdots \\ \overline{J_{vin}} \\ \overline{J_{\rho i1}} \\ \vdots \\ \overline{J_{\rho ij}} \\ \vdots \\ \overline{J_{\rho in}} \end{pmatrix} \Delta \mathrm{d}_\mathrm{i},$$

where the over-line represents the complex conjugate. J_{vij} ($J_{\rho ij}$) is the sensitivity of the ith component of data $\mathrm{d_i}$ with respect to the jth model parameters $\mathrm{m_j}$ of velocity (or density).

Table 10.1
Conjugate gradient algorithm for solving linear system $Ax = b$ [181]

1. Starting from initial vector v_0 and setting the residual vector $r_0 = b - Ax_0$, initial search direction q_0 is set as r_0, i.e., $q_0 = -r_0$.
2. The scalar α_k is computed using the relationship $\alpha_k = \frac{r_k^T r_k}{q_k^T A q_k} = \frac{r_k^T r_k}{q_k^T [J^T (Jq_k)]}$ where k is the number of iterations.
3. The calculation of x_{k+1} is expressed as $x_{k+1} = x_k + \alpha_k q_k$.
4. Then, the next residual vector r_{k+1} is given by $r_{k+1} = r_k - \alpha_k A q_k = r_k - \alpha_k [J^T (Jq_k)]$.
5. The next step is to compute β_k, which is used to find the next search direction $\beta_k = \frac{r_{k+1}^T r_{k+1}}{r_k^T r_k}$.
6. The next search direction q_{k+1} is obtained by using this scalar β_k, $q_{k+1} = -r_{k+1} + \beta_k q_k$.
7. Then, the next scale α_{k+1} is computed by using the obtained search direction q_{k+1} and this process is repeated.

In Equation (10.4), the Hessian is associated with the descent direction Δm. The explicit calculation of the Hessian or the inverse Hessian is very difficult for a large-scale inverse problem. Due to the linearity in this equation, the descent direction can be solved by a matrix-free conjugate gradient method [149], as shown in Table 10.1. In this algorithm, instead of direct calculation of the inverse Hessian, the Hessian-vector products can be used to obtain the descent direction Δm. The Hessian-vector products can be calculated by using two kernel-vector products. First, the kernel-vector products Jq are computed by using the row decomposition

$$Jq = \begin{pmatrix} J_{v11} \cdots J_{v1j} \cdots J_{v1n} \cdots J_{\rho 11} \cdots J_{\rho 1j} \cdots J_{\rho 1n} \\ \vdots \quad \ddots \quad \vdots \quad \ddots \quad \vdots \quad \vdots \quad \ddots \quad \vdots \quad \ddots \quad \vdots \\ J_{vi1} \cdots J_{vij} \cdots J_{vin} \cdots J_{\rho i1} \cdots J_{\rho ij} \cdots J_{\rho in} \\ \vdots \quad \ddots \quad \vdots \quad \ddots \quad \vdots \quad \vdots \quad \ddots \quad \vdots \quad \ddots \quad \vdots \\ J_{vl1} \cdots J_{vlj} \cdots J_{vln} \cdots J_{\rho l1} \cdots J_{\rho lj} \cdots J_{\rho ln} \end{pmatrix} \begin{pmatrix} q_{v1} \\ \vdots \\ q_{vj} \\ \vdots \\ q_{vn} \\ q_{\rho 1} \\ \vdots \\ q_{\rho j} \\ \vdots \\ q_{\rho n} \end{pmatrix} = \begin{pmatrix} J_{v1} q_v + J_{\rho 1} q_\rho \\ \vdots \\ J_{vi} q_v + J_{\rho i} q_\rho \\ \vdots \\ J_{vl} q_v + J_{\rho l} q_\rho \end{pmatrix}, \tag{10.15}$$

where q is an arbitrary vector. Second, the kernel-vector products $J^\dagger J q_k$ can be obtained based on Equation (10.14).

Therefore, the truncated Gauss-Newton method used in multi-parameter FWI consists of two nested loops. First, the external loop is the model iterative update based on Equation (10.9). Second, in order to obtain the descent direct, the internal loop is the iteration of the linear system following Equation (10.4) by using the

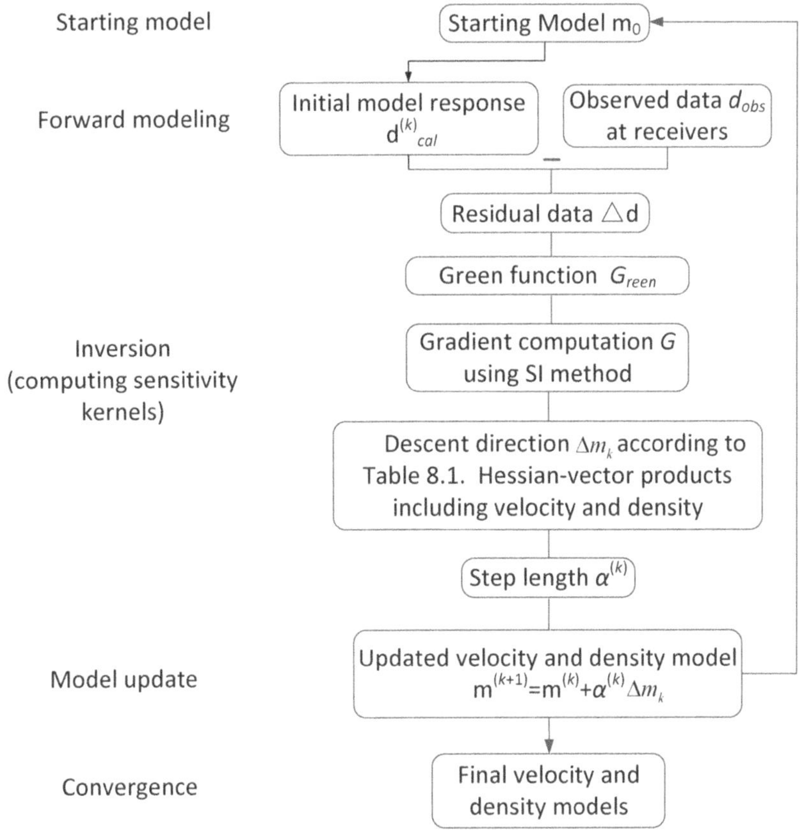

Figure 10.1 The workflow of the truncated Gauss-Newton FWI [181].

conjugate gradient algorithm shown in Table 10.1. The workflow of the truncated Gauss-Newton FWI is shown in Figure 10.1. For simplicity, the frequency loop and the iteration loop for searching the descent direction are omitted in the workflow. In general, there are three steps in this workflow: solving the forward modeling, calculating sensitivity kernels, and updating the model. The overall process is similar to Figure 4.2 shown in Chapter 4. The differences are the methods for obtaining the gradient and the descent direction: the mono-parameter FWI uses the AS method to calculate the gradient and the steepest descent algorithm to search the descent direction, and the multi-parameter FWI uses the SI approach to obtain the gradient and the matrix-free conjugate gradient algorithm to find the descent direction.

In Figure 10.1, the computation of sensitivity kernels is an important step in multi-parameter FWI. The sensitivity kernel represents the sensitivity of the data with respect to the model parameters. Based on the same model perturbation, a larger

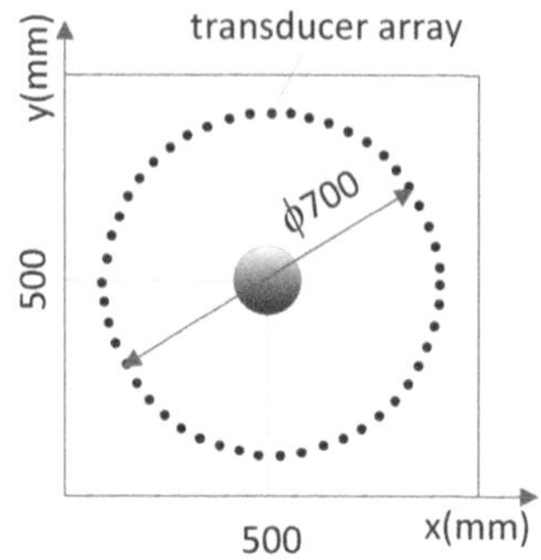

Figure 10.2 One central inclusion model with a 64 element circular array [181].

sensitivity kernel for a parameter denotes a larger sensitivity of the data to this parameter. The frequency-domain sensitivity kernels of velocity and density in the homogeneous background can be given by [266]

$$\begin{aligned} J_v(x) &= \frac{-2\omega^2}{\rho_0(x)v_0^2(x)} G_{reen0}(r,\omega;x)p_0(x,\omega) \\ J_\rho(x) &= -\frac{\omega^2}{\rho_0(x)v_0^2(x)} G_{reen0}(r,\omega;x)p_0(x,\omega) + \frac{1}{\rho_0(x)} \nabla G_{reen0}(r,\omega;x) \cdot \nabla p_0(x,\omega), \end{aligned} \tag{10.16}$$

where $p_0(x,\omega)$ is the pressure wave-field from a transmitter position in the background model and $G_{reen0}(r,\omega;x)$ is the Green's function at the receiver position in the background medium.

10.3 NUMERICAL RESULTS

10.3.1 NUMERICAL MODEL

2D FE simulations carried out by Abaqus are performed in aluminum blocks (density = 2700 kg/m^3, longitudinal wave velocity = 6198 m/s and Poisson's ratio = 0.33) with the dimension of 1000 mm × 1000 mm. Four-node plane strain elements with the size of 1 mm are used in the mesh, which can ensure more than 30 elements per wavelength [98]. A 100 mm diameter inclusion of simultaneously varying velocity of 5843 m/s and density of 2130 kg/m^3 (bulk modulus = 54.9 GPa) located in the center of the plane is modeled, as shown in Figure 10.2. The absorbing regions around the plane are used to avoid boundary reflections. The signals are monitored by a 700 mm circular array with 64 nodes as transducers, as shown in Figure 10.2. The input signal is a 2 cycle Hanning-windowed tone-burst centered at the frequency of 200 kHz.

FD simulations are performed in a 2D space domain, with the same size as in FE models. The plane is meshed by grid points with 2 mm spacing, which satisfies the calculation accuracy requirement of at least 4 grid points per shortest wavelength [87]. The absorbing areas with the same width as in FE models are used around the plane edges. Omni-directional pressure waves are generated and monitored with the same array configuration as presented in the FE setup.

10.3.2 DATA PROCESSING

Since the multi-parameter FWI is based on the FD acoustic model, the scattered signals from the FE elastic model need to be calibrated in order to consider deviations between the two models. The calibration factor is defined as the ratio of the required frequency domain data of the homogeneous background model (without defects) by the FD method and the FE method on the homogeneous structure, as discussed in Section 4.3.3. A contour plot of the typical displacement signal measured over receivers will be shown in Section 10.4.2.

10.3.3 RECONSTRUCTIONS

The model is performed with a circular inclusion with different velocities and densities situated in the center of a 64 node circular array, as shown in Figures 10.3(a) and (c). The diameter of the inclusion is 100 mm (around 3.4λ at 200 kHz). Eight consecutive frequencies between 40 kHz and 180 kHz with the same interval are used for the inversion. A maximum of 40 external iterations for the frequency loop and 20 internal iterations for searching the descent direction are performed for each frequency. The homogeneous backgrounds of velocity and density are used as the starting models at the lowest frequency. The obtained results in the final iterations at the previous frequency are chosen as the starting models for the next higher frequency.

Figures 10.3(b) and (d) outline the velocity and density reconstruction of a single central inclusion based on the elastic model at 180 kHz. It can be clearly seen from the figure that the velocity reconstruction can capture almost all details of the inclusion; however, the reconstructed density image has a smaller size than the original model. The reason is that the velocity can govern both travel times and amplitudes of the waves at all angles, while the density has a dominant effect on the reflection amplitudes at smaller angles. This is consistent with the fact that the wave-field scattered by the velocity perturbation has an isotropic diffraction pattern, which corresponds to the virtual source behaving as a pressure source; amplitudes of the wave-field scattered by density decrease progressively from transmission (low frequencies in the spatial frequency domain) to reflection (high frequencies in the spatial frequency domain) (see Figure 2.1), which corresponds to the virtual source behaving as a vertical force pointing upward in the excitation direction (reflection part) [68, 150]. The velocity reconstruction can be obtained from all acquisitions due to no filtering applied on the velocity gradient by the virtual pressure source; however, the short-to-intermediate wavelengths can be achieved in the density reconstruction because the vertical force virtual source applies the high-pass filtering on the density gradient [150]. Also, the parameter of velocity has a dominant effect on the data residual; thus this parameter is reconstructed before the parameter of density. The accuracy of velocity reconstruction can also affect density reconstruction. The remaining discrepancy is attributed to the difference between acoustic scattering and elastic

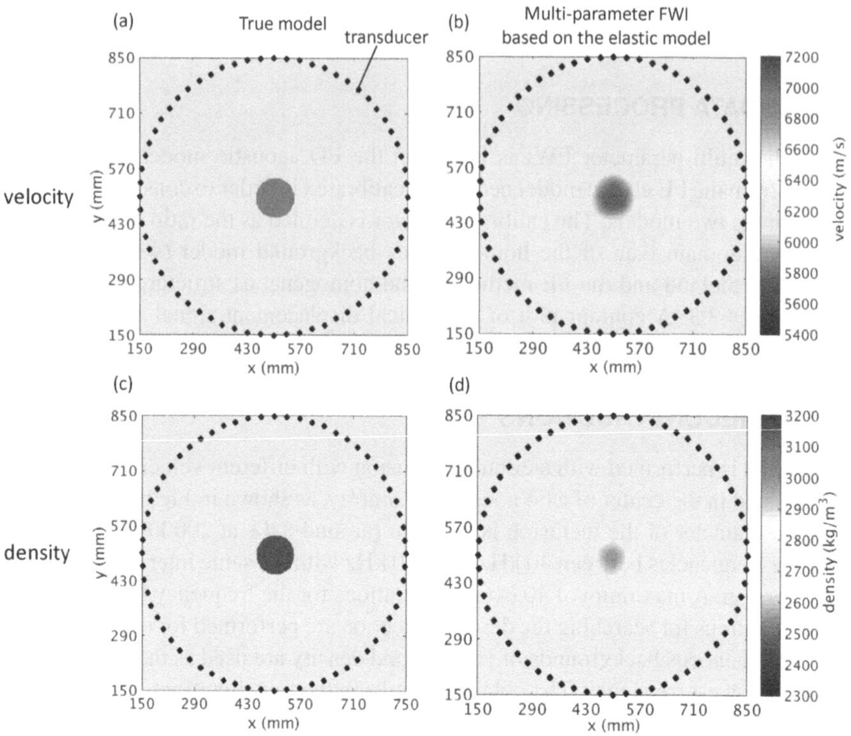

Figure 10.3 Velocity and density reconstruction of a single central inclusion with a diameter of 100 mm in an aluminum block from the elastic model. The images along different rows represent the velocity reconstruction and the density reconstruction. (a) The true model of velocity and (b) the velocity reconstruction; (c) the true model of density and (d) the density reconstruction [181].

scattering since the acoustic inversion approach is used to reconstruct the elastic data.

The cross sections of reconstructions from Figure 10.3 along the horizontal line in the elastic model are shown in Figures 10.4(a) and (b). It can be seen that the velocity reconstruction is slightly overestimated and the density reconstruction is underestimated. When the 2D acoustic inversion is applied to the data from the 2D acoustic forward model, the cross sections of reconstructions of velocity and density are presented in Figures 10.4(c) and (d). It can be seen that the accuracy of the reconstructions still suffers, mainly reflecting the trade-off effects. Compared with reconstructions using the data from the elastic model shown in Figures 10.4(a) and (b), reconstructions from the acoustic model shown in Figures 10.4(c) and (d) are slightly better.

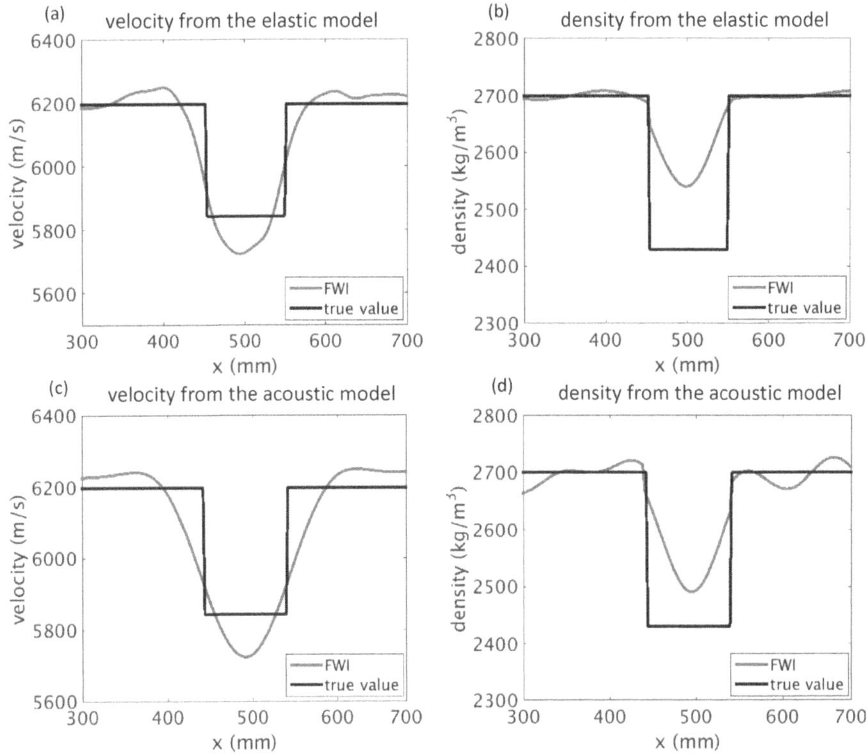

Figure 10.4 Cross sections of velocity and density reconstruction of a single central inclusion along y = 500 mm from (a)–(b) the elastic model and (c)–(d) the acoustic model [181].

10.4 EXPERIMENTS

10.4.1 EXPERIMENTAL SETUP

The experimental setup is shown in Figure 10.5(a). The experiment is performed on a diffusion-bonded cylinder of brass (outer)/stainless steel (inner), with an outer diameter of 100 mm, an inner diameter of 10 mm, and a height of 100 mm. Compared with sizes used in numerical models, the dimensions are scaled down in the ratio of around one to ten in the experiment due to easier manufacturing and lower cost. The material properties are shown in Table 10.2. A disk-shaped piezoelectric transducer (PZT5, Sinoceramics Inc., China) with a diameter of 5 mm and a thickness of 1 mm is attached to the surface of the cylinder using the conductive epoxy. A 5-cycle Hanning-windowed tone-burst signal centered at 1 MHz was generated by RITEC RAM-5000 high power ultrasonic pulser (Ritec Inc., USA). It needs to be noted that the size of the numerical model is scaled from 1000 mm to 100 mm (i.e., a scaling factor of 10), and the frequency changes from 200 kHz (used in

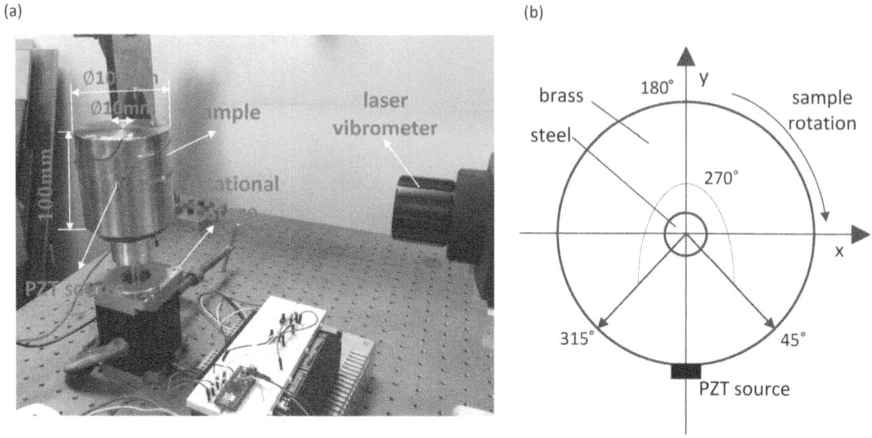

Figure 10.5 (a) The experimental setup, including the PZT source, laser vibrometer, the sample and rotational stage, and (b) the geometry of the experimental setup from the top view [181].

Table 10.2
Material properties [181]

Material	Young's modulus (GPa)	Poisson's ratio	Density (kg/m^3)
Aluminum	70	0.33	2700
Stainless steel	210	0.287	7850
Brass	110	0.34	8700

Section 10.3.1) to 2 MHz. The lower central frequency of 1 MHz is used in the experiment compared with the corresponding frequency of 2 MHz used in the scaling sizes of numerical models. The reasons are: (1) at the lower frequency, the minimum number of transducers in the circular array is reduced because it is not practical to have too many measured points in this experiment; and (2) the attenuation of the ultrasound at lower frequency is less in the brass/steel cylinder. The sample is mounted on a computer-controlled rotational stage, where the positions of the source and receiver are fixed. The displacements are measured by a Polytec OFV-505 laser vibrometer normal to the cylinder axis. The source and receiver are located in the same plane. By rotating the computer-controlled stage, measurements are taken along three quarters of the circle with 123 receiver points, excluding the small range of angles close to the PZT source, as shown in Figure 10.5(b). Since the sample is axisymmetric, the measurements are only taken from a single source and then copied to generate a signal matrix of 164×123.

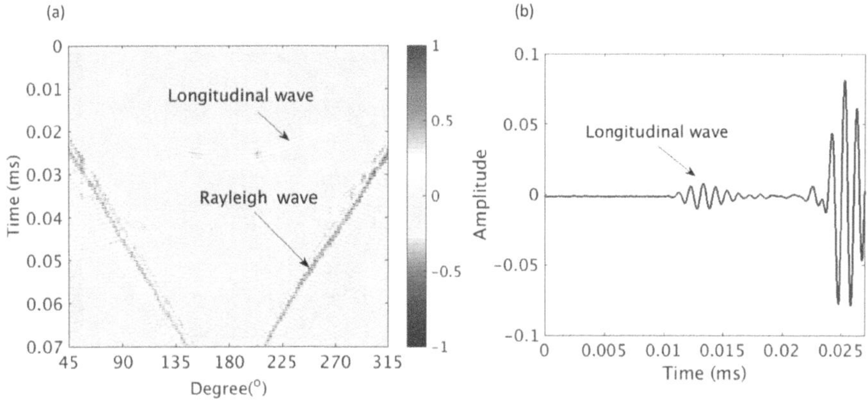

Figure 10.6 (a) Displacement field for the brass/steel cylinder and (b) a typical time trace from one receiver [181].

10.4.2 DATA PROCESSING

The experimental signals are post-processed before using them in the image algorithm. Figures 10.6(a) and (b) show a contour plot of the normal displacement field for all receivers and a typical signal measured by the laser vibrometer in the experiment. It can be seen from Figure 10.6(a) that the direct longitudinal wave has a curved moveout and the Rayleigh wave propagates around the cylinder with a linear moveout. The direct longitudinal wave is separated from signals by removing the slower Rayleigh wave and the unwanted components with the gating function (see Figure 10.6(b)). All signals after time gating are calibrated before using them in the inversion based on a similar method introduced in Section 10.3.2.

10.4.3 EXPERIMENTAL RESULTS

Figures 10.7(a) and (c) show the true velocity and density models of a single circular inclusion located at the center of a 164-transducer circular array with a diameter of 100 mm. The steel inclusion has a diameter of 10 mm (around 1.7λ at 1 MHz). The inversion is carried out by eight consecutive frequencies, ranging from 650 kHz to 1 MHz with the same interval. The homogeneous background models at the lowest frequency are used as the starting models for velocity and density, with 40 external iterations for the frequency loop and 20 internal iterations for searching the descent direction at each frequency.

Figures 10.7(b) and (d) present the reconstructed velocity and density of the inclusion from the experiment by the multi-parameter FWI. It can be seen from Figure 10.7(b) that the diameter of velocity reconstruction of the inclusion is smaller than the true velocity size. The overall shape of the density reconstruction of the inclusion is also narrow, and the contrast of the inclusion is underestimated. This is mainly due

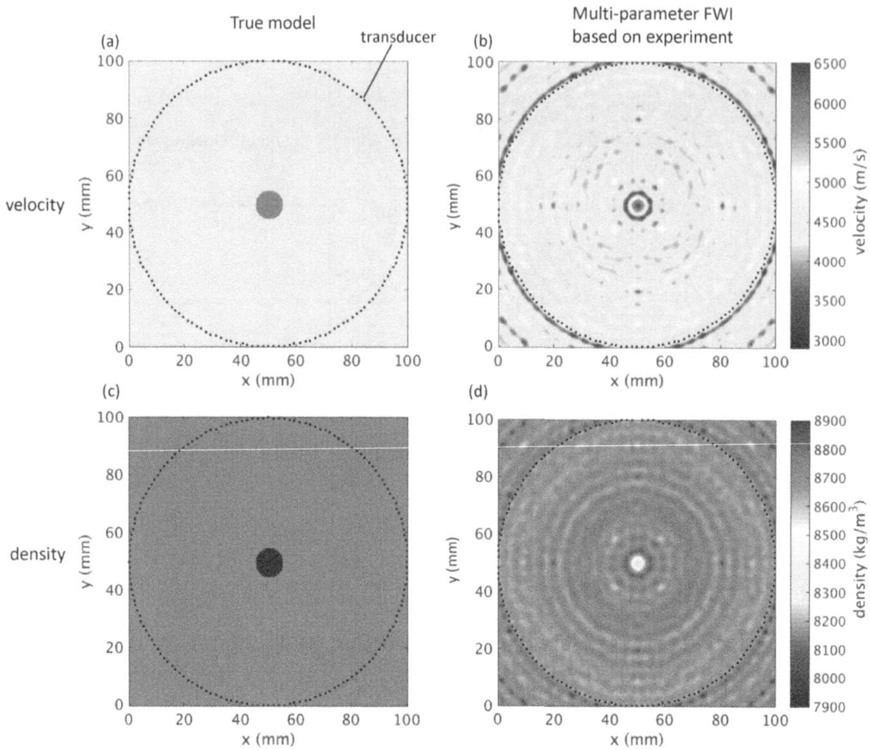

Figure 10.7 The velocity and density mapping of the brass/steel cylinder at 1 MHz in the experiment. Two rows represent velocity and density reconstructions from the multi-parameter FWI [181].

to the limited view configuration of the ultrasonic tomography system, constraints from the 2D acoustic model applied to the realistic elastic data, and trade-off effects between velocity and density.

It needs to be mentioned that regularization was not used in this study. In the presence of noise in the measured data, appropriate regularization introduced in the misfit function deals with the ill-posedness of the FWI algorithm. The regularization, which helps to alleviate the noise effect, can influence the behavior of the truncated Gauss-Newton method by the damping of the Hessian. More details on the Hessian used to improve the descent direction and the smoothing regularization constraint can be found in [6, 135, 266].

Noticeable artifacts are observed both imaging in the experiment. In addition to the reasons already discussed, the excitation wave-fields cannot be ideally omni-directional, which cause the amplitude and phase of signals to vary with the exciting angle. In the calibration, the signals without the defect are synthetically obtained, which implies that the mismatch between the synthesized and the true signals leads to

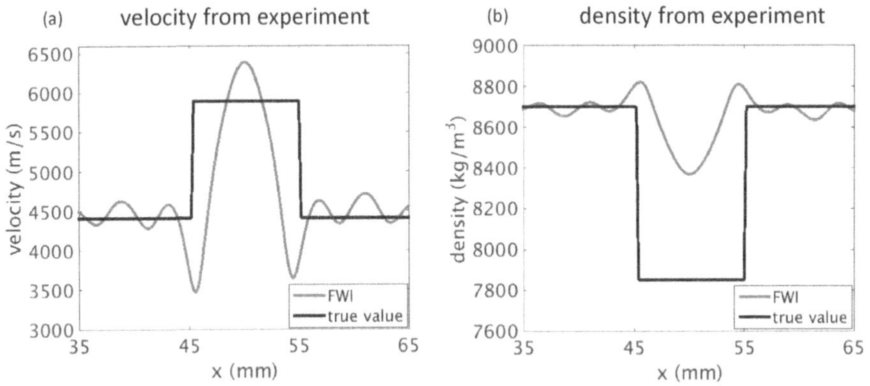

Figure 10.8 Cross sections of velocity and density reconstruction along y = 50 mm from Figure 10.7 [181].

errors in amplitude and phase in the inversion. In addition, measuring position error, slight anisotropy of the sample material, and the length of the gate function can affect the reconstruction quality. In this experiment, the axisymmetric ring artifacts are expected due to the 164 × 123 matrix of source-receiver combinations by repeating only 123 measurements from one source [97].

The cross sections of Figure 10.7 along y = 50 mm are shown in Figures 10.8(a) and (b). These confirm that the size of the reconstructed velocity is smaller than the original one, and the FWI tomography tends to overestimate the velocity reconstruction. Furthermore, the size of the obtained density of the inclusion is narrower than the true one, and the lowest value of the density reconstruction is underestimated.

10.5 SUMMARY

In this chapter, an ultrasonic tomography approach based on the acoustic multi-parameter FWI using the truncated Gauss-Newton method is developed for velocity and density reconstructions of metal parts. An accurate estimation of the inverse Hessian is very important in the multi-parameter FWI. Instead of directly calculating the inverse Hessian, it can be taken into account through a matrix-free conjugate gradient solution of the Gauss-Newton normal equation. With this consideration, trade-off effects between velocity and density are effectively alleviated, and the reconstructions of velocity and density are feasible. Centered inclusions are used to demonstrate the multi-parameter FWI algorithm via simulation and experiment. The results show that the velocity reconstruction aligns well with the true model, and the estimated density is also accurately achieved. This method, which can be extended to guided wave tomography, could prove useful for quality control in the aviation and automotive industries.

10.6 EXERCISES

1. What are the common methods of measuring the density of metal components? How to quantitatively reconstruct the density of a structure containing multiple layers of different metals?
2. How to use ultrasonic tomography to realize the reconstruction of metal density and Young's modulus?
3. What challenges does multi-parameter FWI face compared to mono-parameter FWI?
4. How to mitigate the effect of cross-talk in the multi-parameter inversion?
5. What are the advantages and disadvantages of using the truncated Gauss-Newton method in this chapter?
6. What is the difference between the frequency-domain 2D acoustic wave equation in the isotropic heterogeneous media and isotropic homogeneous media?
7. How to use the conjugate gradient algorithm to solve the linear system?
8. Please draw the workflow of the truncated Gauss-Newton FWI.
9. Why is the reconstructed density image smaller in size than the original model?
10. Please analyze why velocity reconstruction is slightly overestimated and density reconstruction is underestimated.
11. What is Rayleigh wave? What is the difference between the longitudinal wave and the Rayleigh wave?
12. Please analyze the reasons for the difference between the simulation results and the experimental results in this chapter.
13. Please analyze how to improve the density reconstruction.
14. Please analyze the advantages and disadvantages of using the frequency-domain 2D acoustic wave equation and the frequency-domain 2D elastic wave equation in the isotropic heterogeneous media.

References

1. G. B. Aarfken and H. J. Weber. *Mathematical Methods for Physicists*. Academic Press, London, 2001.
2. F. Abe, K. Osakada, M. Shiomi, K. Uematsu, and M. Matsumuto. The manufacturing of hard tools from metallic powders by selective laser melting. *J. Mater. Process. Technol.*, 111(1-3):210–213, 2001.
3. P. R. Amestoy, A. Guermouche, J. Y. L'Excellent, and S. Pralet. Hybridscheduling for the parallel solution of linear systems. *Parallel Comput.*, 32:136–156, 2006.
4. T. Amine, J. W. Newkirk, and F. Liou. Investigation of effect of process parameters on multilayer builds by direct metal deposition. *Appl. Therm. Eng.*, 73:500–511, 2014.
5. K. K. Amireddy, K. Balasubramaniam, and P. Rajagopal. Deep subwavelength ultrasonic imaging using optimized holey structured metamaterials. *Sci. Rep.*, 7(1):1–8, 2017.
6. A. Y. Anagaw and M. D. Sacchi. Comparison of multifrequency selection strategies for simultaneous-source full-waveform inversion. *Geophysics*, 79(5):R165–R181, 2014.
7. M. A. Anastasio and X. Pan. A new reconstruction approach for reflection mode diffraction tomography. *IEEE Trans. Image Process.*, 9(7):1262–1271, 2000.
8. A. Andersen. Algebraic reconstruction in CT from limited views. *IEEE Trans. Med. Imag.*, 8(1):50–55, 1989.
9. J. E. Anderson, J. R. Krebs, and D. Hinkley. Source near the free-surface boundary: pitfalls for elastic finite-difference seismic simulation and multigrid waveform inversion. *70th Annual Conference & Exhibition, EAGE, Extended Abstracts*, page WO10, 2008.
10. G. Aranguren, P. M. Monje, V. Cokonaj, et al. Ultrasonic wave-based structural health monitoring embedded instrument. *Rev. Sci. Instrum.*, 84(12):125106, 2013.
11. ASTM International. Standard test method for apparent density of metal powders and related compounds using the arnold meter (astm b703-17), 2021.
12. B. A. Auld. *Acoustic Fields and Waves in Solids*. Krieger, Malabar, FL, 1990.
13. M. Azimi and A. C. Kak. Distortion in diffraction tomography caused by multiple scattering. *IEEE Trans. Med. Imaging*, MI-2(4):176–195, 1983.
14. A. Bandyopadhyay and B. Heer. Additive manufacturing of multi-material structures. *Mater. Sci. Eng. R Reports*, 129:1–16, 2018.
15. H. Baumgartl, J. Tomas, R. Buettner, and M. Merkel. A deep learning-based model for defect detection in laser-powder bed fusion using in-situ thermographic monitoring. *Prog. Addit. Manuf.*, 5:277–285, 2020.
16. P. Belanger. *Feasibility of thickness mapping using ultrasonic guided waves*. PhD thesis, Univ. of Imperial College, London, 2009.
17. P. Belanger and P. Cawley. Lamb wave tomography to evaluate the maximum depth of corrosion patches. *Rev. Prog. Quant. Nondestruct Eval.*, 27:1290–1297, 2008.
18. P. Belanger, P. Cawley, and F. Simonetti. Guided wave diffraction tomography within the Born approximation. *IEEE Trans. Ultrason. Ferroelectr. Freq. Control*, 57(6):1405–1418, 2010.
19. J. P. Berenger. A perfectly matched layer for absorption of electromagnetic waves. *J. Comput. Phys.*, 114:185–200, 1994.

20. P. B. Bertrand, R. A. Levine, E. M. Isselbacher, and P. M. Vandervoort. Fact or artifact in two-dimensional echocardiography: avoiding misdiagnosis and missed diagnosis. *J. Am. Soc. Echocardiogr.*, 29:381–391, 2063.
21. S. E. Bobbin, J. W. Wagner, and R. C. Cammarata. Determination of the flexural modulus of thin films from measurement of the first arrival of the symmetric Lamb wave. *Appl. Phys. Lett.*, 59(13):1544–1546, 1991.
22. M. Born and E. Wolf. *Principles of Optics*. Cambridge University Press, Cambridge, UK, 1999.
23. A. J. Brath, F. Simonetti, and P. B. Nagy. Acoustic formulation of elastic guided wave propagation and scattering in curved tubular structures. *IEEE Trans. Ultrason. Ferroelectr. Freq. Control*, 61(5):815–829, 2014.
24. A. J. Brath, F. Simonetti, P. B. Nagy, and G. Instanes. Experimental validation of a fast forward model for guided wave tomography of pipe elbows. *IEEE Trans. Ultrason. Ferroelectr. Freq. Control*, 64(5):859–871, 2017.
25. A. J. Brath, F. Simonetti, P. B. Nagy, and G. Instanes. Guided wave tomography of pipe bends. *IEEE Trans. Ultrason. Ferroelectr. Freq. Control*, 64(5):847–858, 2017.
26. L. M. Brekhovskikh and V. Goncharov. *Mechanics of Continua and Wave Dynamics*. Springer-Verlag, 1985.
27. R. Brossier, S. Operto, and J. Virieux. Seismic imaging of the complex on shore structures by 2D elastic frequency-domain full-waveform inversion. *Geophysics*, 74(6):WCC105–WCC118, 2009.
28. A. Bulleti, P. Giannelli, M. Calzolai, and L. Capineri. An integrated acousto/ultrasonic structural health monitoring system for composite pressure vessels. *IEEE Trans. Ultrason. Ferroelectr. Freq. Control*, 63(6):864–873, 2016.
29. C. Bunks, F. M. Saleck, S. Zaleski, and G. Chavent. Multiscale seismic waveform inversion. *Geophysics*, 60(5):1457–1473, 1995.
30. J. V. Burke, F. E. Curtis, A. S. Lewis, et al. Gradient sampling methods for nonsmooth optimization. *Numerical nonsmooth optimization: State of the art algorithms*, pages 201–225, 2020.
31. James V Burke, Adrian S Lewis, and Michael L Overton. A robust gradient sampling algorithm for nonsmooth, nonconvex optimization. *SIAM J. Optim.*, 15(3):751–779, 2005.
32. B. E. Carroll, R. A. Otis, J. P. Borgonia, et al. Functionally graded material of 304L stainless steel and inconel 625 fabricated by directed energy deposition: Characterization and thermodynamic modeling. *Acta Mater.*, 108(15):46–54, 2016.
33. M. Castaings and C. Bacon. Finite element modeling of torsional wave modes along pipes with absorbing materials. *J. Acoust. Soc. Am.*, 119(6):3741–3751, 2006.
34. P. Cawley. Long range inspection of structures using low frequency ultrasound. *Structural Damage Assessment Using Advanced Signal Processing Procedures*, pages 1–17, 1997.
35. F. B. Cegla and J. Allin. Wireless ultrasonic thickness monitoring at elevated temperatures. *Mater. Eval.*, 69:26–31, 2011.
36. F. B. Cegla, P. Cawley, J Allin, and J. Davies. High-temperature (>500° C) wall thickness monitoring using dry-coupled ultrasonic waveguide transducers. *IEEE Trans. Ultrason. Ferroelectr. Freq. Control.*, 58:158–167, 2010.
37. F. B. Cegla, A. Rohde, and M. Veidt. Analytical prediction and experimental measurement for mode conversion and scattering of plate waves at non-symmetric circular blind holes in isotropic plates. *Wave Motion*, 45(3):162–177, 2008.

38. A. Chabot, N. Laroche, E. Carcreff, et al. Towards defect monitoring for metallic additive manufacturing components using phased array ultrasonic testing. *J. Intell. Manuf.*, 31:11911201, 2020.
39. F. C. Chen and W. C. Chew. Experimental verification of super resolution in nonlinear inverse scattering. *Appl. Phys. Lett.*, 72(23):3080–3082, 1998.
40. P. Chen, T. H. Jordan, and L. Zhao. Full three-dimensional tomography: A comparison between the scattering-integral and adjoint-wavefield methods. *Geophys. J. Int.*, 170:175–181, 2007.
41. W. C. Chew and Y. M. Wang. Reconstruction of two-dimensional permittivity distribution using the distorted Born iterative method. *IEEE Trans. Med. Imag.*, 9(2):218–225, 1990.
42. Y. Cho. Estimation of ultrasonic guided wave mode conversion in a plate with thickness variation. *IEEE Trans. Ultrason. Ferroelectr. Freq. Control*, 47(3):591–603, 2000.
43. Y.-H. Chueh, C Wei, X. Zhang, and L. Li. Integrated laser-based powder bed fusion and fused filament fabrication for three-dimensional printing of hybrid metal/polymer objects. *Addit. Manuf.*, 31:100928, 2020.
44. T. Clarke, P. Cawley, P. D. Wilcox, and A. J. Croxford. Evaluation of the damage detection capability of a sparse-array guided-wave SHM system applied to a complex structure under varying thermal conditions. *IEEE Trans. Ultraso.n Ferroelectr. Freq. Control*, 56:2666–2678, 2009.
45. R. Clayton and B. Engquist. Absorbing boundary conditions for acoustic and elastic wave equations. *Bull. Seism. Soc. Am.*, 67(6):1529–1540, 1977.
46. A. C. Cobb and J. L. Fisher. Flaw depth sizing using guided waves. *AIP Conf. Proc.*, 1706:030013, 2016.
47. D. Colton and A. Kirsch. A simple method for solving inverse scattering problems in the resonance region. *Inverse Probl.*, 12(4):383–393, 1996.
48. D. Colton and R. Kress. *Inverse Acoustic and Electromagnetic Scattering Theory*. Springer, Heidelberg, 1998.
49. A. J. Croxford, P. D. Wilcox, B. W. Drinkwater, et al. Strategies for guided-wave structural health monitoring. *Proc. Math. Phys. Eng. Sci. A*, 463(2087):2961–2981, 2007.
50. A. Dass and A. Moridi. State of the art in directed energy deposition: From additive manufacturing to materials design. *Coatings*, 9:418, 2019.
51. M. Davison. The ill-conditioned nature of the limited angle tomography problem. *SIAM J. Appl. Math.*, 43(2):428–448, 1983.
52. J. L. Dean, C. Trillo, A. F. Doval, and J. L. Fernandez. Determination of thickness and elastic constants of aluminum plates from full-field wavelength measurements of single-mode narrowband Lamb waves. *J. Acoust. Soc. Am.*, 124(3):1477–1489, 2008.
53. A. Delaney and Y. Bresler. Globally convergent edge-preserving regularized reconstruction: An application to limited-angle tomography. *IEEE Trans. Image Process.*, 7(2):204–221, 1998.
54. A. G. Demir and B. Previtali. Multi-material selective laser melting of Fe/Al-12Si components. *Manuf. Lett.*, 11:8–11, 2017.
55. A. Devaney. Geophysical diffraction tomography. *IEEE Trans. Geosci. Remote Sens.*, 22(1):3–13, 1984.
56. A. P. Dhawan, R. M. Rangayyan, and R. Gordon. Image restoration by Wiener deconvolution in limited-view computed tomography. *Appl. Opt.*, 24:4013–4020, 1985.

57. M. Drozdz, Skelton E., Craster R. V., and Lowe M. J. S. Modeling builk and guided waves in unbounded elastic media using absorbing layers in commercial finite elmeent packages. *AIP Conf. Proc*, 894:87–94, 2007.
58. M. B. Drozdz. *Efficient finite element modelling of ultrasound waves in elastic media*. Imperial College, London, 2008.
59. N. Duric, P. Littrup, L. Poulo, and A. Babkin. Detection of breast cancer with ultrasound tomography: First results with the computed ultrasound risk evaluation (CURE) prototype. *Med. Phys.*, 34:773, 2007.
60. E. Duveneck, P. Milcik, P. M. Bakker, and P. Perkins. Acoustic VTI wave equations and their application for anisotropic reverse-time migration. *78th Annual International Meeting, SEG, Expanded Abstracts*, pages 2186–2190, 2008.
61. P. S. Earle and P. M. Shearer. Characterization of global seismograms using an automatic-picking algorithm. *Bull. Seismol. Soc. Am.*, 84:366–376, 1994.
62. J. W. Eberhard. Ultrasonic tomography for nondestructive evaluation. *Ann. Rev. Mater. Sci.*, 12:1–21, 1982.
63. A. Eremin, E. Glushkov, N. Glushkova, and R. Lammering. Guided wave time-reversal imaging of macroscopic localized inhomogeneities in anisotropic composites. *Struct. Health. Monit.*, 18:1–17, 2019.
64. F. L. Everett and J. Miklowitz. Poisson's ratio at high temperatures. *Journ. Appl. Phys.*, 15:592–598, 1944.
65. P. P. Ewald. Introduction to the dynamical theory of x-ray diffraction. *Acta Crystallogr. Sect. A*, 25(1):103–108, 1969.
66. C. R. Farrar and K. Worden. *Structural health monitoring: a machine learning perspective*. John Wiley & Sons, 2012.
67. M. J. Fleming. *Far-field super resolution imaging*. PhD thesis, Univ. of Imperial College, 2008.
68. E. Forgues and G. Lambare. Parameterization study for acoustic and elastic ray plus Born inversion. *J. Seism. Explor.*, 6:253–278, 1997.
69. P. Fromme, P. D. Wilcox, M. J. S. Lowe, and P. Cawley. On the development and testing of a guided ultrasonic wave array for structural integrity monitoring. *IEEE Trans. Ultrason. Ferroelectr. Freq. Control*, 53:777–785, 2006.
70. H. Gao, Y. Shi, and J. L. Rose. Guided wave tomography on an aircraft wing with leave in place sensors. *AIP Conf. Proc.*, 760:1788–1794, 2004.
71. W. Gao, Y. Zhang, D. Ramanujan, et al. The status, challenges, and future of additive manufacturing in engineering. *Comput. Aided Des.*, 69:65–89, 2015.
72. P. Gardner, R. Fuentes, N. Dervilis, et al. Machine learning at the interface of structural health monitoring and non-destructive evaluation. *Philos. Trans. Royal Soc. A*, 378(2182):20190581, 2020.
73. I. Geir, M. A. Toppe, and P. B. Nagy. Method and system of measuring material loss from a solid structure. *U.S. Patent No. 8,225,665*, 2012.
74. R. M. German. *Powder Metallurgy Science*. NJ: Metal Powder Industries Federation, Princeton, 1994.
75. Ali Gholami, Hossein S Aghamiry, and Stephane Operto. Extended-space full-waveform inversion in the time domain with the augmented Lagrangian method. *Geophysics*, 87(1):R63–R77, 2022.
76. I. Gibson, D. Rosen, and B. Stucker. *Directed energy deposition processes. Additive manufacturing technologies: 3D printing, rapid prototyping, and direct digital manufacturing*. New York: Springer New York, 2015.

77. H. J. Gong, K. Rafi, H. F. Gu, T. Starr, and B. Stucker. Analysis of defect generation in Ti–6Al–4V parts made using powder bed fusion additive manufacturing processes. *Addit. Manuf.*, 1-4:87–98, 2014.
78. J. W. Goodman. *Fourier Optics*. McGraw-Hill, London, 1996.
79. K. F. Graff. *Wave motion in elastic solids*. Dover, New York, 1975.
80. K. M. Hanson. *Image Recovery: Theory and Application*. Academic, New York, 1987.
81. J. He, J. Rao, J. D. Fleming, et al. Numerical ultrasonic full waveform inversion (FWI) for complex structures in coupled 2D solid/fluid media. *Smart Mater. Struct.*, 30:085044, 2021.
82. J. C. Heigel, P. Michaleris, and E. W. Reutzel. Thermo-mechanical model development and validation of directed energy deposition additive manufacturing of Ti-6Al-4V. *Addit. Manuf.*, 5:9–19, 2015.
83. A. J. Hesford and W. C. Chew. Fast inverse scattering solutions using the distorted Born iterative method and the multilevel fast multipole algorithm. *J. Acoust. Soc. Am.*, 128(2):679–690, 2010.
84. A. H. Hielscher, A. D. Klose, and K. M. Hanson. Gradient-based iterative image reconstruction scheme for time-resolved optical tomography. *IEEE Trans. Med. Imag.*, 18(3):262–271, 1999.
85. F. Honarvar and A. Varvani-Farahani. A review of ultrasonic testing applications in additive manufacturing: Defect evaluation, material characterization, and process control. *Ultrasonics*, 108:106227, 2020.
86. S. Howard, J. Tani, H. Arnold, and W. Sachse. Ultrasonic characterization of porosity in powder metals. *NDE of Microstructure for Process Control ed H.N.G. Wadley, ASM, PA*, pages 81–89, 1985.
87. B. Hustedt, S. Operto, and J. Virieux. Mixed-grid and staggered-grid finite-difference methods for frequency-domain acoustic wave modelling. *Geophys. J. Int.*, 157:1269–1296, 2004.
88. W. B. Hutchinson, K. Ushioda, and G. Runnsjö. Anisotropy of tensile behaviour in a duplex stainless steel sheet. *Mater. Sci. and Technol.*, 1:728–731, 1985.
89. P. Huthwaite. *Quantitative imaging with mechanical waves*. PhD thesis, Univ. of Imperial College, 2009.
90. P. Huthwaite. Evaluation of inversion approaches for guided wave thickness mapping. *Proc. R. Soc. A.*, 470:20140063, 2014.
91. P. Huthwaite. Guided wave tomography with an improved scattering model. *Proc. Royal Soc. A*, 472(2195):20160643, 2016.
92. P. Huthwaite. Improving accuracy through density correction in guided wave tomography. *Proc. R. Soc. A.*, 472:20150832, 2016.
93. P. Huthwaite, M. J. S. Lowe, and P. Cawley. Guided wave tomography performance analysis. *AIP Conference Proceedings*, 1706:020021, 2016.
94. P. Huthwaite, R. Ribichini, P. Cawley, and M. J. S. Lowe. Mode selection for corrosion detection in pipes and vessels via guided wave tomography. *IEEE Trans. Ultrason. Ferroelectr. Freq. Control*, 60(6):1165–1177, 2013.
95. P. Huthwaite and M. Seher. Robust helical path separation for thickness mapping of pipes by guided wave tomography. *IEEE Trans. Ultrason. Ferroelectr. Freq. Control*, 62(5):927–938, 2015.
96. P. Huthwaite and F. Simonetti. High-resolution imaging without iteration: a fast and robust method for breast ultrasound tomography. *J. Acoust. Soc. Am.*, 130(3):1721–1734, 2011.

97. P. Huthwaite and F. Simonetti. High-resolution guided wave tomography. *Wave Motion*, 50(5):979–993, 2013.
98. P. Huthwaite, F. Simonetti, and M. J. S. Lowe. On the convergence of finite element scattering simulations. *Rev. Prog. Quant. Nondestruct Eval.*, 29:65–72, 2010.
99. P. Huthwaite, A. A. Zwiebel, and F. Simonetti. A new regularization technique for limited-view sound-speed imaging. *IEEE Trans. Ultrason. Ferroelectr. Freq. Control*, 60(3):603–613, 2013.
100. G. Instanes, A. O. Pedersen, and A. Brath. Ultrasonic computerized tomography for continuous monitoring of corrosion and erosion damage in pipelines. *CORROSION 2015 NACE International*, 2015.
101. G. Instanes, M. Toppe, B. Lakshminarayan, and P. B. Nagy. Corrosion and erosion monitoring of pipes by an ultrasonic guided wave method. *Advanced Ultrasonic Methods for Material and Structure Inspection*, pages 115–157, 2007.
102. C. H. Jo, C. Shin, and J. H. Suh. An optimal 9-point, finite-difference, frequency-space 2-D scalar extrapolator. *Geophysics*, 61(2):529–537, 1996.
103. A. C. Kak and M. Slaney. *Principles of Computerized Tomographic Reconstruction*. IEEE Press, New York, 1998.
104. N. Kang, X. Lin, M. El Mansori, et al. On the effect of the thermal cycle during the directed energy deposition application to the in-situ production of a Ti-Mo alloy functionally graded structure. *Addit. Manuf.*, 31:100911, 2020.
105. M. R. Karim and A. K. Mal. Inversion of leaky Lamb wave data by simplex algorithm. *J. Acoust. Soc. Am.*, 88(1):482–491, 1990.
106. S. Kessler, S. Spearing, and C. Soutis. Damage detection in composite materials using lamb wave methods. *Smart Mater. Struct.*, 11:269–278, 2002.
107. L. E. Kinsler and A. R. Frey. *Fundamentals of Acoustics*. John Wiley and Sons, New York, 1962.
108. A. Kirsch. Characterization of the shape of a scattering obstacle using the spectral data of the far field operator. *Inverse Probl.*, 14:1489–1512, 1998.
109. A. Kumar and W. Arnold. High resolution in non-destructive testing: A review. *J. Appl. Phys.*, 132(10):100901, 2022.
110. N. Kurgan. Effect of porosity and density on the mechanical and microstructural properties of sintered 316L stainless steel implant materials. *Mater. Des.*, 55:235–241, 2014.
111. H. Lamb. On waves in an elastic plate. *Proc. R. Soc. London, Ser. A*, 93(648):114–128, 1917.
112. S. LaRoque, E. Sidky, and X. Pan. Accurate image reconstruction from few-view and limited-angle data in diffraction tomography. *J. Opt. Soc. Am.*, 25(7):1772–1782, 2008.
113. S. Laureti, D. A. Hutchins, L. Astolfi, et al. Trapped air metamaterial concept for ultrasonic sub-wavelength imaging in water. *Sci. Rep.*, 10(1):1–12, 2020.
114. M. C. Lawrence, M. A. Jaffer, and B. T Sewell. The application of the maximum entropy method to electron microscopic tomography. *Ultramicroscopy*, 31(3):285–301, 1989.
115. C. A. C. Leckey, K. R. Wheeler, V. N. Hafiychuk, H. Hafiychuk, and A. Dogan. Simulation of guided-wave ultrasound propagation in composite laminates: Benchmark comparisons of numerical codes and experiment. *Ultrasonics*, 84:187–200, 2018.
116. M. Lemistre and D. Balageas. Structural health monitoring system based on diffracted lamb wave analysis by multiresolution processing. *Smart Mater. Struct.*, 10(3):504–511, 2001.

117. J. Li and J. L. Rose. Natural beam focusing of non-axisymmetric guided waves in large-diameter pipes. *Ultrasonics*, 44:35–45, 2006.
118. L. Li, W. U. H. Syed, and A. J. Pinkerton. Rapid additive manufacturing of functionally graded structures using simultaneous wire and powder laser deposition. *Virtual Phys. Prototyp.*, 1(4):217–225, 2006.
119. M. Li and Z. Feng. Accurate Young's modulus measurement based on Rayleigh wave velocity and empirical Poisson's ratio. *Rev. Sci. Instrum.*, 87:075111, 2016.
120. S. Li, M. Jackowski, D. Dione, et al. Refraction corrected transmission ultrasound computed tomography for application in breast imaging. *Med. Phys.*, 37:2233–2246, 2010.
121. S. Li, K. Mueller, and M. et al. Jackowski. Fast marching method to correct for refraction in ultrasound computed tomography. *In 3rd IEEE Int. Symp. on Biomedical Imaging: Nano to Macro*, pages 896–899, 2006.
122. G. Liu and J. Qu. Guided circumferential waves in a circular annulus. *J. Appl. Mech.*, 65(2):424–430, 1998.
123. Y. Liu, J. Yang, B. Chi, and L. Dong. An improved scattering-integral approach for frequency-domain full waveform inversion. *Geophys. J. Int.*, 202:1827–1842, 2015.
124. Mathias Louboutin and Felix J Herrmann. Extending the search space of time-domain adjoint-state fwi with randomized implicit time shifts. In *79th EAGE Conference and Exhibition 2017*, pages 1–5. European Association of Geoscientists & Engineers, 2017.
125. M. J. S. Lowe. Matrix techniques for modelling ultrasonic waves in multilayered media. *IEEE Trans. Ultrason. Ferroelectr. Freq. Control*, 42:525–542, 1995.
126. C. Lu, J. Lin, W. Chew, and G. Otto. Image reconstruction with acoustic measurement using distorted Born iteration method. *Ultrason. Imaging*, 18:140–156, 1996.
127. R. D. Luke and R. Potthast. The no response test—a sampling method for inverse scattering problems. *SIAM J. Appl. Math.*, 63(4):1292–1312, 2003.
128. R. M. Mahamood and E. T. Akinlabi. Laser metal deposition of functionally graded Ti6Al4V/TiC. *Mater. Des.*, 84:402–410, 2015.
129. L. Maio, V. Memmelo, F. Ricci, N. D. Boffa, E. Monaco, and R. Pecora. Ultrasonic wave propagation in composite laminates by numerical simulation. *Compos. Struct.*, 121:64–74, 2015.
130. E. V. Malyarenko and M. K. Hinders. Ultrasonic Lamb wave diffraction tomography. *Ultrasonics*, 39(4):269–281, 2001.
131. M. Malyarenko and M. Hinders. Fan beam and double crosshole lamb wave tomography for mapping flaws in aging aircraft structures. *J. Acoust. Soc. Am.*, 108(4):1631–1639, 2000.
132. K. J. Marfurt. Accuracy of finite-difference and finite-element modeling of the scalar and elastic wave-equations. *Geophysics*, 49(5):533–549, 1984.
133. L. Mazeika, R. Kazys, R. Raisutis, and R. Sliteris. Ultrasonic guided wave tomography for the inspection of the fuel tanks floor. *Int. J. Mater. Prod. Technol.*, 41(1-4):128–139, 2011.
134. G. McMechan, J. Harris, and L. Anderson. Cross-hole tomography for strongly variable media with applications to scale model data. *Bull. Seismol. Soc. Am.*, 77(6):1945–1960, 1987.
135. L. Metivier, F. Bretaudeau, R. Brossier, S. Operto, and J. Virieux. Full waveform inversion and the truncated Newton method: Quantitative imaging of complex subsurface structures. *Geophysics*, 62:1353–1375, 2014.

136. L. Metivier and R. Brossier. The seiscope optimization toolbox: A large-scale nonlinear optimization library based on reverse communication. *Geophysics*, 81:F11–F25, 2016.
137. L. Metivier, R. Brossier, S. Operto, and J. Virieux. Acoustic multiparameter FWI for the reconstruction of P-wave velocity, density and attenuation: Preconditioned truncated Newton approach. *85th Annual International Meeting, SEG, Expanded Abstracts*, pages 1198–1203, 2015.
138. Ludovic Metivier and Romain Brossier. Receiver-extension strategy for time-domain full-waveform inversion using a relocalization approach. *Geophysics*, 87(1):R13–R33, 2022.
139. N. Metropolis, A. W. Rosenbluth, and M. N. Rosenbluth. Equation of state calculations by fast computing machines. *J. Chem. Phys.*, 21(6):1087–1091, 1953.
140. C. Millon, A. Vanhoye, A. F. Obaton, and J. D. Penot. Development of laser ultrasonics inspection for online monitoring of additive manufacturing. *Weld. world*, 62:653–661, 2018.
141. R. D. Mindlin. Influence of rotator inertia and shear on flexural motions of isotropic, elastic plate. *J. Appl. Mech.*, 18:31–38, 1951.
142. P. Mora. Nonlinear two-dimensional elastic inversion of multioffset seismic data. *Geophysics*, 52(9):1211–1228, 1987.
143. L. Moreau, M. Caleap, A. Velichko, and P. D. Wilcox. Scattering of guided waves by through-thickness cavities with irregular shapes. *Wave Motion*, 48(7):586–602, 2011.
144. L. Moreau, A. J. Hunter, A. Velichko, and P. D. Wilcox. 3-D reconstruction of subwavelength scatterers from the measurement of scattered fields in elastic waveguides. *IEEE Trans. Ultrason. Ferroelectr. Freq. Control*, 61(11):1864–1879, 2014.
145. B. I. S. Murat, P. Khalili, and P. Fromme. Scattering of guided waves at delaminations in composite plates. *J. Acoust. Soc. Am.*, 139(6):3044–3052, 2004.
146. F. Natterer and F. Wübbeling. *Mathematical methods in image reconstruction*. Society for Industrial and Applied Mathematics, 2001.
147. C. T. Ng. *Guided Wave Scattering Analysis and Damage Diagnosis of Advanced Structural Materials*. PhD thesis, Univ. of Queensland, 2010.
148. L. T. Nguyen and R. T. Modrak. Ultrasonic wavefield inversion and migration in complex heterogeneous structures: 2d numerical imaging and nondestructive testing experiments. *Ultrasonics*, 82:357–370, 2018.
149. J. Nocedal and S. J. Wright. *Numerical optimization (2nd ed.): Springer Science + Business Media*. Springer Series in Operations Research and Financial Engineering, LLC, 2006.
150. S. Operto, Y. Gholami, V. Prieux, et al. A guided tour of multiparameter full waveform inversion for multicomponent data: from theory to practice. *The Leading Edge*, 32(9):1040–1054, 2013.
151. S. Operto, J. Virieux, J. X. Dessa, and G. Pascal. Crustal seismic imaging from multifold ocean bottom seismometer data by frequency domain full waveform tomography: Application to the eastern Nankai trough. *J. Geophys. Res.*, 111:B09306, 2006.
152. S. Operto, J. Virieux, A. Ribodetti, and J. E. Anderson. Finite-difference frequency-domain modeling of viscoacoustic wave propagation in 2D tilted transversely isotropic (TTI) media. *Geophysics*, 74(5):T75–T95, 2009.
153. O. Oyelola, P. Crawforth, R. M'Saoubi, and A. T. Clare. Machining of functionally graded Ti6Al4V/WC produced by directed energy deposition. *Addit. Manuf.*, 24:20–29, 2018.

154. J. P. Panakkal, J. K. Ghosh, and P. R. Roy. Use of ultrasonic velocity for measurement of density of sintered fuel pellets. *J. Phys. D: Appl. Phys.*, 17:1791–1795, 1984.
155. S. Pant, J. Laliberte, M. Martinez, B. Rocha, and D. Ancrum. Effects of composite lamina properties on fundamental Lamb wave mode dispersion characteristics. *Compos. Struct.*, 124:236–252, 2015.
156. E. P. Papadakis. *The measurement of ultrasonic velocity, Physical Acoustics*, volume XIX. Academic Press, New York, 1990.
157. B. Pavlakovic, , M. J. S. Lowe, D. Alleyne, and P. Cawley. Disperse: A general purpose program for creating dispersion curves. *Rev. Prog. Quant. Nondestruct Eval.*, 16:185–192, 1997.
158. M. Persson, D. Bone, and H. Elmqvist. Total variation norm for three-dimensional iterative reconstruction in limited view angle tomography. *Phys. Med. Biol.*, 46(3):853–866, 2001.
159. PiezoMechanik GmbH. Piezo-mechanics: An introduction, 2025.
160. R. E. Plessix. A review of the adjoint-state method for computing the gradient of a functional with geophysical applications. *Geophys. J. Int.*, 167:495–503, 2006.
161. R. Potthast. *Point Sources and Multipoles in Inverse Scattering Theory*. Chemical Rubber Company Press, Boca Raton, FL, 2001.
162. C. Prada, J. L. Thomas, and M. Fink. Decomposition of the time reversal operator: detection and selective focusing on two scatterers. *J. Acoust. Soc. Am.*, 99(4):2067–2076, 1996.
163. R. G. Pratt. Seismic waveform inversion in the frequency domain, Part 1: Theory and verification in a physical scale model. *Geophysics*, 64(3):888–901, 1999.
164. R. G. Pratt, C. Shin, and G. J. Hicks. Gauss-newton and full newton methods in frequency-space seismic waveform inversion. *Geophys. J. Int.*, 133:341–362, 1998.
165. R. G. Pratt and M. H. Worthington. Inverse theory applied to multisource cross-hole tomography. Part 1: acoustic wave-equation method. *Geophys. Prosp.*, 38:287–310, 1990.
166. G. T. Pudipeddi and Ng C. T. Effect of central and non-central frequency commponents on the quality of damage imaging. *J. Civil Sstruct. Health Monit*, 8:49–61, 2018.
167. O. Putkis, R. P. Dalton, and A. J. Croxford. The anisotropic propagation of ultrasonic guided waves in composite materials and implications for practical applications. *Ultrasonics*, 65:390–399, 2016.
168. P. Rajagopal, M. Drozdz, E. Skelton, M. J. S. Lowe, and R. Craster. On the use of absorbing layers to simulate the propagation of elastic waves in unbounded isotropic media using commercially available finite element packages. *NDT & E. Int.*, 51:30–40, 2012.
169. J Rao. *Quantitative Imaging Using Ultrasonic Tomography Based on Full Waveform Inversion*. PhD thesis, Nanyang Technol. Univ., 2018.
170. J. Rao, M. Ratassepp, and Z. Fan. Guided wave tomography based on full waveform inversion. *IEEE Trans. Ultrason. Ferroelectr. Freq. Control*, 63(5):737–745, 2016.
171. J. Rao, M. Ratassepp, and Z. Fan. Limited-view ultrasonic guided wave tomography using an adaptive regularization method. *J. Appl. Phys.*, 120(19):194902, 2016.
172. J. Rao, M. Ratassepp, and Z. Fan. Investigation of the reconstruction accuracy of guided wave tomography using full waveform inversion. *J. Sound Vib.*, 400:317–328, 2017.
173. J. Rao, M. Ratassepp, and Z. Fan. On-line corrosion monitoring of plate structures based on guided wave tomography using piezoelectric sensors. *Sensors*, 17(12), 2882, 2017.

174. J. Rao, M. Ratassepp, and Z. Fan. Quantification of thickness loss in a liquid-loaded plate using ultrasonic guided wave tomography. *Smart Mater. Struct.*, 26:125017, 2017.
175. J. Rao, M. Ratassepp, and Z. Fan. Quantification of thickness loss in liquid-loaded pipes based on guided wave tomography. *Proceedings of the 7th Asia–Pacific Workshop on Structural Health Monitoring*, pages 12–15, 2018.
176. J. Rao, A. Saini, J. Yang, M. Ratassepp, and Z. Fan. Ultrasonic imaging of irregularly shaped notches based on elastic reverse time migration. *NDT & E. Int.*, 107:102135, 2019.
177. J. Rao, S. L. Sing, J. C. W. Lim, W. Y. Yeong, J. Yang, Z. Fan, and P. Hazell. Detection and characterisation of defects in directed energy deposited multi-material components using full waveform inversion and reverse time migration. *Virtual and Physical Prototyping*, 17(4):1047–1057, 2022.
178. J. Rao, J. L. Wang, S. Kollmannsbergera, et al. Point cloud-based elastic reverse time migration for ultrasonic imaging of components with vertical surfaces. *Mech. Syst. Signal Process.*, 163:108144, 2022.
179. J. Rao, F. Yang, H. Mo, et al. Quantitative reconstruction of defects in multi-layered bonded composites using fully convolutional network-based ultrasonic inversion. *J. Sound Vib.*, 542:117418, 2023.
180. J. Rao, J. Yang, J. He, M. Huang, and E. Rank. Elastic least-squares reverse-time migration with density variation for flaw imaging in heterogeneous structures. *Smart Mater. Struct.*, 29:035017, 2020.
181. J. Rao, J. Yang, M. Ratassepp, and Z. Fan. Multi-parameter reconstruction of velocity and density using ultrasonic tomography based on full waveform inversion. *Ultrasonics*, 101(106004), 2020.
182. C. O. Rasgado-Moreno and M. Ratassepp. Geodesic equations for guided wave helical path separation for a pipe bend. *Mech. Syst. Signal Process.*, 204(110820), 2023.
183. C. O. Rasgado-Moreno, M. Rist, R. Land, and M. Ratassepp. Acoustic forward modelfor guided wave propagation and scattering in a pipe bend. *Sensors*, 22(486), 2022.
184. C. O. Rasgado-Moreno, M. Rist, R. Land, and M. Ratassepp. Guided wave tomography of pipe bends based on full waveform inversion. *Ultrasonics*, 148(107560), 2025.
185. C. O. Rasgado-Moreno, P. Xu, M. Rist, and M. Ratassepp. Optimising full waveform inversion with inhomogeneous transducers: Parameters and considerations for successful implementation. *NDT & E. Int.*, 149(103265), 2025.
186. M. Ratassepp, M. J. S. Lowe, P. Cawley, and A. Klauson. Scattering of the fundamental shear horizontal mode in a plate when incident at a through-crack aligned in the propagation direction of the mode. *J. Acoust. Soc. Am.*, 124(5):2873–2882, 2008.
187. M. Ratassepp, J. Rao, and Z. Fan. Quantitative imaging of Young's modulus in plates using guided wave tomography. *NDT & E. Int.*, 94:22–30, 2018.
188. M. Ratassepp, J. Rao, X. Yu, and Z. Fan. Modeling the effect of anisotropy in ultrasonic-guided wave tomography. *IEEE Transactions on Ultrasonics, Ferroelectrics, and Frequency Control*, 69(1):330–339, 2021.
189. C. Ravaut, S. Operto, L. Improta, and J. Virieux. Multiscale imaging of complex structures from multifold wide-aperture seismic data by frequency-domain full-waveform tomography: Application to a thrust belt. *Geophys. J. Int.*, 159:1032–1056, 2004.
190. A. Reichardt, R. P. Dillon, J. Paul, et al. Development and characterization of Ti-6Al-4V to 304L stainless steel gradient components fabricated with laser deposition additive manufacturing. *Mater. Des.*, 104:404–413, 2016.

191. M. L. Reis and N. C. Roberty. Maximum entropy algorithms for image reconstruction from projections. *Inverse Probl.*, 8:623–644, 1992.
192. H. Ren, H. Wang, and S. Chen. Least-squares reverse time migration in frequency domain using the adjoint-state method. *J. Geophys. Eng.*, 10:035002, 2013.
193. R. Ribichini, F. Cegla, P. B. Nagy, and P. Cawley. Experimental and numerical evaluation of electromagnetic acoustic transducer performance on steel materials. *NDT & E Int.*, 45:32–38, 2012.
194. P. R Roberge. *Corrosion Inspection and Monitoring*. New Jersey, John Wiley, 2007.
195. W. P. Rogers. Elastic property measurement using Rayleigh-Lamb waves. *Res. Nondestr. Eval.*, 6:185–208, 1995.
196. A. H. Rohde, M. Veidt, L. R. F. Rose, and J. Homer. A computer simulation study of imaging flexural inhomogeneities using plate-wave diffraction tomography. *Ultrasonics*, 48(1):6–15, 2008.
197. J. L. Rose. *Ultrasonic waves in solid media*. Cambridge University Press, Cambridge, 1999.
198. L. R. F. Rose and C. H. Wang. Mindlin plate theory for damage detection: Source solutions. *J. Acoust. Soc. Am.*, 116:154–171, 2004.
199. L. R. F. Rose and C. H. Wang. Mindlin plate theory for damage detection: Imaging of flexural inhomogeneities. *J. Acoust. Soc. Am.*, 127(2):754–763, 2010.
200. D. Royer, D. Clorennec, and C. Prada. Lamb mode spectra versus the Poisson ratio in a free isotropic elastic plate. *J. Acoust. Soc. Am.*, 125(6):3683–3687, 2009.
201. C. K. Rushforth and R. W. Harris. Restoration, resolution and noise. *J. Opt. Soc. Am.*, 58(4):539–545, 1968.
202. A. Saboori, D. Gallo, S. Biamino, P. Fino, and M. Lombardi. An overview of additive manufacturing of titanium components by directed energy deposition: Microstructure and mechanical properties. *Appl. Sci.*, 7(9):883, 2017.
203. Y. Sakumoto, T. Nakazato, and A. Matsuzaki. High-temperature properties of stainless steel for building structures. *J. Struct. Eng.*, 122(4):399–406, 1996.
204. F. Schumacher, W. Friederich, and S. Lamara. A flexible, extendable, modular and computationally efficient approach to scattering-integral based seismic full waveform inversion. *Geophys. J. Int.*, 204:1100–1119, 2016.
205. C. B. Scruby and B. C. Moss. Non-contact ultrasonic measurements on steel at elevated temperatures. *NDT & E. Int.*, 26:177–188, 1993.
206. A. Segura, P. Lanceleur, J. F. De Belleval, J. M. Gherbezza, and F. Lesage. Influence of anisotropy for the characterizartion of internal imperfection in pipes by ultrasonic non-destructive testing. *e-Journal of Nondestructive Testing (NDT)*, 14(5):1–11, 2009.
207. M. Seher, P. Huthwaite, and M. J. S. Lowe. Experimental studies of the inspection of areas with restricted access using A0 lamb wave tomography. *IEEE Trans. Ultrason. Ferroelectr. Freq. Control*, 63:1455–1467, 2016.
208. R. Seidl and E. Rank. Iterative time reversal based flaw identification. *Comput. Math. Appl*, 72:879–892, 2016.
209. K. Shah, I. U. Haq, A. Khan, S. A. Shah, M. Khan, and A. J. Pinkerton. Parametric study of development of Inconel-steel functionally graded materials by laser direct metal depositiong. *Mater. Des.*, 54:531–538, 2014.
210. D. H. Sheen, K. Tuncay, C. E. Baag, and P. J. Ortoleva. Time domain Gauss-Newton seismic waveform inversion in elastic media. *Geophys. J. Int.*, 167:1373–1384, 2006.
211. J. W. C. Sherwood. Elastic wave propagation in a semi-infinite solid medium. *Proc. Phys. Soc*, 71(2):207, 1958.

212. C. Shin, K. Yoon, K. J. Marfurt, and K. Park. Efficient calculation of a partial derivative wavefield using reciprocity for seismic imaging and inversion. *Geophysics*, 66(6):1856–1863, 2001.
213. E. Sidky and X. Pan. Image reconstruction in circular cone-beam computed tomography by constrained, total-variation minimization. *Phys. Med. Biol.*, 53(17):4777–4807, 2008.
214. F. Simonetti. Breaking the resolution limit: A new perspective for imaging in NDT. *Rev. Prog. Quant. Nondestruct. Eval.*, 25:700–707, 2006.
215. F. Simonetti. Localization of pointlike scatterers in solids with subwavelength resolution. *Appl. Phys. Lett.*, 89(9):094105, 2006.
216. F. Simonetti. Multiple scattering: The key to unravel the subwavelength world from the far-field pattern of a scattered wave. *Phys. Rev. E*, 73:036619, 2006.
217. F. Simonetti, M. Fleming, and E A. Marengo. Illustration of the role of multiple scattering in subwavelength imaging from far-field measurements. *J. Opt. Soc. Am. A*, 25(2):292–303, 2008.
218. F. Simonetti, L. Huang, and N. Duric. On the spatial sampling of wave fields with circular ring apertures. *J. Appl. Phys.*, 101(8):083103, 2007.
219. F. Simonetti, L. Huang, N. Duric, and P. Littrup. Diffraction and coherence in breast ultrasound tomography: A study with toroidal array. *Med. Phys.*, 36(7):2955–2965, 2009.
220. L. Sirgue and R. G. Pratt. Efficient waveform inversion and imaging: A strategy for selecting temporal frequencies. *Geophysics*, 69(1):231–248, 2004.
221. M. Slaney, A. C. Kak, and L. E. Larsen. Limitations of imaging with first-order diffraction tomography. *IEEE Trans. Microwave Theory Tech.*, 32(8):860–874, 1984.
222. D. Slepian and H. O. Pollak. Prolate spheroidal wave functions, Fourier analysis and uncertainty-I. *Bell Syst. Tech. J.*, 40:43–63, 1961.
223. R. J. Smith, M. Hirsch, R. Patel, et al. Spatially resolved acoustic spectroscopy for selective laser melting. *J. Mater. Process. Technol.*, 236:93–102, 2016.
224. T. Sol, S. Hayun, D. Noiman, et al. Nondestructive ultrasonic evaluation of additively manufactured AlSi10Mg samples. *Addit. Manuf.*, 22:700–707, 2018.
225. H. Song and Y. Yang. Super-resolution visualization of subwavelength defects via deep learning-enhanced ultrasonic beamforming: A proof-of-principle study. *NDT & E. Int.*, 116:102344, 2020.
226. A. B. Spierings, M. Schneider, and R. Eggenberger. Comparison of density measure techniques for additive manufactured metallic parts. *Rapid Prototyp. J.*, 7:380–386, 2011.
227. G. Sposito, P. Cawley, and P. B. Nagy. Potential drop mapping for the monitoring of corrosion or erosion. *NDT & E Int.*, 43:394–402, 2010.
228. P. R. Sreeraj, S. K. Mishra, and P. K. Singh. A review on non-destructive evaluation and characterization of additively manufactured components. *Prog. Addit. Manuf.*, pages 1–24, 2021.
229. I. Stekl and R. G. Pratt. Accurate viscoelastic modeling by frequency-domain finite differences using rotated operators. *Geophysics*, 63(5):1779–1794, 1998.
230. Y. Sung, W. Choi, C. Fang-Yen, and K. Badizadegan. Optical diffraction tomography for high resolution live cell imaging. *Opt. Express*, 17(1):266–277, 2009.
231. E. M. T. Takougang and A. J. Calvert. Application of waveform tomography to marine seismic reflection data from the Queen Charlotte Basin of western Canada. *Geophysics*, 76(2):B55–B70, 2011.

232. S. Tammas-Williams and I. Todd. Design for additive manufacturing with site-specific properties in metals and alloys. *Scr. Mater.*, 135:105–110, 2017.
233. Z. E. E. Tan, J. H. L. Pang, J. Kaminski, and H. Pepin. Characterisation of porosity, density, and microstructure of directed energy deposited stainless steel AISI 316L. *Addit. Manuf.*, 25:286–296, 2019.
234. A. Tarantola. Inversion of seismic reflection data in the acoustic approximation. *Geophysics*, 49(8):1259–1266, 1984.
235. A. Tarantola. A strategy for nonlinear elastic inversion of seismic reflection data. *Geophysics*, 51(10):1893–1903, 1986.
236. A. Tarantola. *Inverse Problem Theory: Methods for Data Fitting and Parameter Estimation*. Elsevier, 1987.
237. L. Thomsen. Weak elastic anisotropy. *Geophysics*, 51(10):1954–1966, 1986.
238. W. T. Thomson. Transmission of elastic waves through a stratified solid medium. *J. Appl. Phys.*, 21:89–93, 1950.
239. M. N. Toksoz and D. H. Johnston. *Seismic wave attenuation*. Society of Exploration Geophysicists, Tulsa, Okla, 1981.
240. G. Toraldo di Francia. Degrees of freedom of an image. *J. Opt. Soc. Am.*, 59(7):799–804, 1969.
241. I. Tsvankin. Anisotropic parameters and P-wave velocity for orthorombic media. *Geophysics*, 64(4):1292–1309, 1997.
242. A. Uriondo, M. Esperon-Miguez, and S. Perinpanayagam. The present and future of additive manufacturing in the aerospace sector: A review of important aspects. *J. Aerosp. Eng.*, 229(11):2132–2147, 2015.
243. M Vaezi, S. Chianrabutra, B. Mellor, and S. Yang. Multiple material additive manufacturing-Part 1: a review. *Virtual Phys. Prototyp.*, 8(1):19–50, 2013.
244. A. Velichko and P. D. Wilcox. Excitation and scattering of guided waves: relationships between solutions for plates and pipes. *J. Acoust. Soc. Am.*, 125(6):3623–3631, 2009.
245. A. Velichko and P. D. Wilcox. A generalized approach for efficient finite element modeling of elastodynamic scattering in two and three dimensions. *J. Acoust. Soc. Am.*, 128(3):1004–1014, 2010.
246. C. Vemula and A. N. Norris. Flexural wave propagation and scattering on thin plates using Mindlin theory. *Wave Motion*, 26:1–12, 1997.
247. D. Vigh, E. W. Starr, and J. Kapoor. Developing earth model with full waveform inversion. *The Leading Edge*, 28:432–435, 2009.
248. S. Vinogradov, C. Duffer, and G. M. Light. Magnetostrictive sensing probes for guided wave testing of high temperature pipes. *Mater. Eval.*, 72:803–811, 2014.
249. J. Virieux and S. Opertor. An overview of full-waveform inversion in exploration geophysics. *Geophysics*, 74(6):WCC1–WCC26, 2009.
250. A. Volker and J. Bloom. Experimental results of guided wave travel time tomography. *Rev. Prog. Quant. Nondestruct. Eval.*, 30:215–222, 2011.
251. A. Volker and T. van Zon. Guided wave travel time tomography for bends. *AIP Conference Proceedings*, 1511(1):737–744, 2013.
252. T. Wandowski, P. Malinowski, and W. M. Ostachowicz. Damage detection with concentrated configurations of piezoelectric transducers. *Smart Mater. Struct.*, 20:025002, 2011.
253. C. H. Wang and Chang F. K. Scattering of plate waves by a cylindrical inhomogeneity. *J. Sound Vib.*, 282:429–451, 2005.

254. D. J. Wang, W. F. Zhang, X. Y. Wang, and B. Sun. Lamb-wave-based tomographic imaging techniques for hole-edge corrosion monitoring in plate structures. *Materials*, 9:916, 2016.
255. L. Wang and F. G. Yuan. Group velocity and characteristic wave curves of Lamb waves in composites: modeling and experiments. *Compos. Sci. Tech.*, 67:1370–1384, 2007.
256. X. Wang, X. Lu, and J. Tang. Damage detection using piezoelectric transducers and the lamb wave approach: I. System analysis. *Smart Mater. Struct.*, 17:025033, 2008.
257. Y. M. Wang and W. C. Chew. An iterative solution of the two-dimensional electromagnetic inverse scattering problem. *Int. J. Imaging Syst. Technol.*, 1:100–108, 1989.
258. M. Warner, A. Ratcliffe, T. Nangoo, et al. Anisotropic 3D full-waveform inversion. *Geophysics*, 78:R59–R80, 2013.
259. C. L. Willey and F. Simonetti. A two-dimensional analysis of the sensitivity of a pulse first break to wave speed contrast on a scale below the resolution length of ray tomography. *J. Acoust. Soc. Am.*, 139(6):3145–3158, 2016.
260. P. R. Williamson. A guide to the limits of resolution imposed by scattering in ray tomography. *Geophysics*, 56(2):202–207, 1991.
261. E. Wolf and J. T. Foley. Do evanescent waves contribute to the far field? *Opt. Lett.*, 23(1):16–18, 1998.
262. K. Worden and J. M. Dulieu-Barton. An overview of intelligent fault detection in systems and structures. *Struct. Health Monit.*, 3(1):85–98, 2004.
263. I. Yadroitsev and I. Yadroitsava. Evaluation of residual stress in stainless steel 316L and Ti6Al4V samples produced by selective laser melting. *Virtual Phys. Prototyp.*, 10(2):67–76, 2015.
264. T. Yamazaki. Development of a hybrid multi-tasking machine tool: integration of additive manufacturing technology with CNC machining. *Procedia CIPR*, 42:81–86, 2016.
265. F. Yan, R. L. Royer, and J. L. Rose. Ultrasonic guided wave imaging techniques in structural health monitoring. *J. Intell. Mater. Syst. Struct.*, 21:377–384, 2010.
266. J. Yang, Y. Liu, and L. Dong. Simultaneous estimation of velocity and density in acoustic multiparameter full-waveform inversion using an improved scattering-integral approach. *Geophysics*, 81(6):R399–R415, 2016.
267. J. Z. Yang, Y. E. Li, Y. Liu, et al. Mitigating the cycle-skipping of full-waveform inversion by random gradient sampling. *Geophysics*, 85(6):R493–R507, 2020.
268. S. Yang and J. R. G. Evans. A multi-component powder dispensing system for three dimensional functional gradients. *Mater. Sci. Eng. A.*, 379:351–359, 2004.
269. H. Yeung, Z. Yang, and L. Yan. A meltpool prediction based scan strategy for powder bed fusion additive manufacturing. *Addit. Manuf.*, 35:101383, 2020.
270. J. W. Young and J. C. Bertrand. Multiple scattering by two cylinders. *J. Acoust. Soc. Am.*, 58(6):1190–1195, 1975.
271. H. Zhang, C. Thurber, and C. Rowe. Automatic P-wave arrival detection and picking with multiscale wavelet analysis for single-component recordings. *Bull. Seismol. Soc. Am.*, 84:1904–1912, 2003.
272. R. Zhang, M. Wan, and W. Cao. Parameter measurement of thin elastic layers using low-frequency multi-mode ultrasonic Lamb waves. *IEEE Trans. Instrum. Meas.*, 50(5):1397–1403, 2001.
273. Y. Zhang, W. Jarosinski, Y.-G. Jung, and J. Zhang. Additive manufacturing processes and equipment. *Additive Manufacturing, Materials, Processes, Quantifications and Applications*, pages 39–51, 2018.

274. X. Zhao, J. L. Rose, and H. D. Gao. Determination of density distribution in ferrous powder compacts using ultrasonic tomography. *IEEE Trans. Ultrason. Ferroelectr. Freq. Control*, 53(2):360–369, 2006.
275. X. Zhao, R. L. Royer, S. E. Owens, and J. L. Rose. Ultrasonic Lamb wave tomography in structural health monitoring. *Smart Mater. Struct.*, 20(10):105002, 2011.
276. H. Zhou, G. Zhang, and R. Bloor. An anisotropic acoustic wave equation for modeling and migration in 2D TTI media. *76th Annual International Meeting, SEG, Expanded Abstracts*, pages 194–198, 2006.
277. H. W. Zhou, H. Hu, Z. Zou, et al. Reverse time migration: A prospect of seismic imaging methodology. *Earth-Sci. Rev.*, 179:207–227, 2018.

Index

For Product Safety Concerns and Information please contact our EU representative GPSR@taylorandfrancis.com Taylor & Francis Verlag GmbH, Kaufingerstraße 24, 80331 München, Germany